Solutions Manual to Accompany

Simulation and the Monte Carlo Method

BICENTENNIAL
1807
⊕ WILEY
2007
BICENTENNIAL

THE WILEY BICENTENNIAL–KNOWLEDGE FOR GENERATIONS

*E*ach generation has its unique needs and aspirations. When Charles Wiley first opened his small printing shop in lower Manhattan in 1807, it was a generation of boundless potential searching for an identity. And we were there, helping to define a new American literary tradition. Over half a century later, in the midst of the Second Industrial Revolution, it was a generation focused on building the future. Once again, we were there, supplying the critical scientific, technical, and engineering knowledge that helped frame the world. Throughout the 20th Century, and into the new millennium, nations began to reach out beyond their own borders and a new international community was born. Wiley was there, expanding its operations around the world to enable a global exchange of ideas, opinions, and know-how.

For 200 years, Wiley has been an integral part of each generation's journey, enabling the flow of information and understanding necessary to meet their needs and fulfill their aspirations. Today, bold new technologies are changing the way we live and learn. Wiley will be there, providing you the must-have knowledge you need to imagine new worlds, new possibilities, and new opportunities.

Generations come and go, but you can always count on Wiley to provide you the knowledge you need, when and where you need it!

WILLIAM J. PESCE
PRESIDENT AND CHIEF EXECUTIVE OFFICER

PETER BOOTH WILEY
CHAIRMAN OF THE BOARD

Solutions Manual to Accompany

Simulation and the Monte Carlo Method

Second Edition

Dirk P. Kroese

University of Queensland
Department of Mathematics
Brisbane, Australia

Thomas Taimre

University of Queensland
Department of Mathematics
Brisbane, Australia

Zdravko I. Botev

University of Queensland
Department of Mathematics
Brisbane, Australia

Reuven Y. Rubinstein

Technion-Israel Institute of Technology
Faculty of Industrial Engineering and Management
Haifa, Israel

WILEY-INTERSCIENCE
A John Wiley & Sons, Inc., Publication

Published by John Wiley & Sons, Inc., Hoboken, New Jersey.
Published simultaneously in Canada.

For general information on our other products and services or for technical support, please contact our
Customer Care Department within the United States at (800) 762-2974, outside the United States at
(317) 572-3993 or fax (317) 572-4002.

Wiley also publishes its books in a variety of electronic formats. Some content that appears in print may
not be available in electronic format. For information about Wiley products, visit our web site at
www.wiley.com.

Wiley Bicentennial Logo: Richard J. Pacifico

Library of Congress Cataloging-in-Publication Data is available.

ISBN 978-0-470-25879-8

CONTENTS

PREFACE

The only effective way to master the theory and practice of Monte Carlo simulation is through exercises and experiments. For this reason, we (RYR and DPK) included many exercises and algorithms in *Simulation and the Monte Carlo Method, 2nd Edition* (SMCM2), Wiley & Sons, New York, 2007.

This companion volume to SMCM2 is written in the same style and contains a wealth of additional material: worked solutions for the over 180 problems in SMCM2, lots of Matlab algorithms and many illustrations.

Like SMCM2, this solution manual is aimed at a broad audience of students, instructors and researchers in engineering, physical and life sciences, statistics, computer science and mathematics, as well as anyone interested in using Monte Carlo simulation in his or her study or work. One of the main goals of the book is to provide a comprehensive solutions guide to instructors, which will aid student assessment and stimulate further student development. In addition, the book offers a unique complement to SMCM2 for self-study. All too often a stumbling block for learning is the unavailability of worked solutions and actual algorithms.

The problem set includes the major topics in Monte Carlo simulation. Starting with exercises that review various important concepts in probability and optimization, the book covers a wide range of exercises in random variable/process generation, discrete-event simulation, statistical analysis of output data, variance reduction techniques, Markov chain Monte Carlo, simulated annealing, sensitivity analysis, cross-entropy methods, and Monte Carlo counting techniques. The original questions from SMCM2 have been added for easy reference. More difficult sections and exercises are marked with an asterisk ($*$) sign. Our choice of using Matlab was motivated by its ease of use and clarity of syntax.

Reference numbers to SMCM2 are indicated in boldface font. For example, **Definition 1.1.1** refers to the corresponding definition in SMCM2, and **(1.7)** refers to equation (1.7) in SMCM2, whereas Figure 1.1 refers to the first numbered figure in the present book.

DIRK KROESE, THOMAS TAIMRE, ZDRAVKO BOTEV AND REUVEN RUBINSTEIN

Brisbane and Haifa
July, 2007

ACKNOWLEDGMENTS

We thank Gareth Evans for providing the code for Problem 3.8.

This book was supported by the Australian Research Council, under Grants DP056631 and DP055895. Thomas Taimre acknowledges the financial support of the Australian Research Council Centre of Excellence for Mathematics and Statistics of Complex Systems.

DPK, TT, ZIB, RYR

PART I

PROBLEMS

CHAPTER 1

PRELIMINARIES

Probability Theory

1.1   Using the properties of the probability measure in **Definition 1.1.1**:

> A *probability* $\mathbb{P}$ is a rule that assigns a number $0 \leqslant \mathbb{P}(A) \leqslant 1$ to each event A, such that $\mathbb{P}(\Omega) = 1$, and such that for any sequence $A_1, A_2, \ldots$ of disjoint events
>
> $$\mathbb{P}\left(\bigcup_i A_i\right) = \sum_i \mathbb{P}(A_i) \,,$$

prove the following results.

   (a) $\mathbb{P}(A^c) = 1 - \mathbb{P}(A)$.
   (b) $\mathbb{P}(A \cup B) = \mathbb{P}(A) + \mathbb{P}(B) - \mathbb{P}(A \cap B)$.

1.2   Prove the product rule **(1.4)**:

> For any sequence of events $A_1, A_2, \ldots, A_n$,
>
> $$\mathbb{P}(A_1 \cdots A_n) = \mathbb{P}(A_1)\,\mathbb{P}(A_2 \,|\, A_1)\,\mathbb{P}(A_3 \,|\, A_1 A_2) \cdots \mathbb{P}(A_n \,|\, A_1 \cdots A_{n-1}) \,,$$
>
> using the abbreviation $A_1 A_2 \cdots A_k \equiv A_1 \cap A_2 \cap \cdots \cap A_k$.

for the case of three events.

Solutions Manual for SMCM, 2nd Edition. By D.P. Kroese, T. Taimre, Z.I. Botev, and R.Y. Rubinstein
Copyright © 2007 John Wiley & Sons, Inc.

1.3 We draw three balls consecutively from a bowl containing exactly five white and five black balls, without putting them back. What is the probability that all drawn balls will be black?

1.4 Consider the random experiment where we toss a biased coin until heads comes up. Suppose the probability of heads on any one toss is p. Let X be the number of tosses required. Show that $X \sim G(p)$.

1.5 In a room with many people, we ask each person his/her birthday, for example, May 5. Let N be the number of people queried until we get a "duplicate" birthday.
  (a) Calculate $\mathbb{P}(N > n)$, $n = 0, 1, 2, \dots$.
  (b) For which n do we have $\mathbb{P}(N \leqslant n) \geqslant 1/2$?
  (c) Use a computer to calculate $\mathbb{E}[N]$.

1.6 Let X and Y be independent standard normal random variables, and let U and V be random variables that are derived from X and Y via the linear transformation

$$\begin{pmatrix} U \\ V \end{pmatrix} = \begin{pmatrix} \sin \alpha & -\cos \alpha \\ \cos \alpha & \sin \alpha \end{pmatrix} \begin{pmatrix} X \\ Y \end{pmatrix} .$$

  (a) Derive the joint pdf of U and V.
  (b) Show that U and V are independent and standard normally distributed.

1.7 Let $X \sim \text{Exp}(\lambda)$. Show that the *memoryless property* holds:

$$\mathbb{P}(X > t + s \,|\, X > t) = \mathbb{P}(X > s) \quad \text{for all } s, t \geqslant 0 .$$

1.8 Let X_1, X_2, X_3 be independent Bernoulli random variables with success probabilities $1/2, 1/3$, and $1/4$, respectively. Give their conditional joint pdf, given that $X_1 + X_2 + X_3 = 2$.

1.9 Verify the expectations and variances in Table 1.1 below.

Table 1.1 Expectations and variances for some well-known distributions.

Dist.	$\mathbb{E}[X]$	$\text{Var}(X)$	Dist.	$\mathbb{E}[X]$	$\text{Var}(X)$
$\text{Bin}(n, p)$	np	$np(1-p)$	$\text{Gamma}(\alpha, \lambda)$	$\dfrac{\alpha}{\lambda}$	$\dfrac{\alpha}{\lambda^2}$
$G(p)$	$\dfrac{1}{p}$	$\dfrac{1-p}{p^2}$	$N(\mu, \sigma^2)$	μ	σ^2
$\text{Poi}(\lambda)$	λ	λ	$\text{Beta}(\alpha, \beta)$	$\dfrac{\alpha}{\alpha+\beta}$	$\dfrac{\alpha\beta}{(\alpha+\beta)^2(1+\alpha+\beta)}$
$U(\alpha, \beta)$	$\dfrac{\alpha + \beta}{2}$	$\dfrac{(\beta - \alpha)^2}{12}$	$\text{Weib}(\alpha, \lambda)$	$\dfrac{\Gamma(1/\alpha)}{\alpha\lambda}$	$\dfrac{2\Gamma(2/\alpha)}{\alpha} - \left(\dfrac{\Gamma(1/\alpha)}{\alpha\lambda}\right)^2$
$\text{Exp}(\lambda)$	$\dfrac{1}{\lambda}$	$\dfrac{1}{\lambda^2}$			

1.10   Let X and Y have joint density f given by

$$f(x, y) = c\,x\,y, \quad 0 \leqslant y \leqslant x, \quad 0 \leqslant x \leqslant 1 \,.$$

(a) Determine the normalization constant c.
(b) Determine $\mathbb{P}(X + 2Y \leqslant 1)$.

1.11   Let $X \sim \mathsf{Exp}(\lambda)$ and $Y \sim \mathsf{Exp}(\mu)$ be independent. Show that

(a) $\min(X, Y) \sim \mathsf{Exp}(\lambda + \mu)$,

(b) $\mathbb{P}(X < Y \mid \min(X, Y)) = \dfrac{\lambda}{\lambda + \mu}$.

1.12   Verify the properties of variance and covariance in Table 1.2 below.

Table 1.2   Properties of variance and covariance.

1	$\mathrm{Var}(X) = \mathbb{E}[X^2] - (\mathbb{E}[X])^2$
2	$\mathrm{Var}(aX + b) = a^2 \mathrm{Var}(X)$
3	$\mathrm{Cov}(X, Y) = \mathbb{E}[XY] - \mathbb{E}[X]\,\mathbb{E}[Y]$
4	$\mathrm{Cov}(X, Y) = \mathrm{Cov}(Y, X)$
5	$\mathrm{Cov}(aX + bY, Z) = a\,\mathrm{Cov}(X, Z) + b\,\mathrm{Cov}(Y, Z)$
6	$\mathrm{Cov}(X, X) = \mathrm{Var}(X)$
7	$\mathrm{Var}(X + Y) = \mathrm{Var}(X) + \mathrm{Var}(Y) + 2\,\mathrm{Cov}(X, Y)$
8	X and Y indep. $\Longrightarrow \mathrm{Cov}(X, Y) = 0$

1.13   Show that the correlation coefficient always lies between -1 and 1. (Hint: use the fact that the variance of $aX + Y$ is always nonnegative, for any a.)

1.14   Consider **Examples 1.1–1.2.** Define X as the function that assigns the number $x_1 + \cdots + x_n$ to each outcome $\omega = (x_1, \ldots, x_n)$. The event that there are exactly k heads in n throws can be written as

$$\{\omega \in \Omega : X(\omega) = k\} \,.$$

If we abbreviate this to $\{X = k\}$ and further abbreviate $\mathbb{P}(\{X = k\})$ to $\mathbb{P}(X = k)$, then we obtain exactly **(1.7)**. Verify that one can always view random variables in this way, that is, as real-valued functions on Ω, and that probabilities such as $\mathbb{P}(X \leqslant x)$ should be interpreted as $\mathbb{P}(\{\omega \in \Omega : X(\omega) \leqslant x\})$.

1.15   Show that

$$\mathrm{Var}\left(\sum_{i=1}^{n} X_i\right) = \sum_{i=1}^{n} \mathrm{Var}(X_i) + 2 \sum_{i<j} \mathrm{Cov}(X_i, X_j) \,.$$

1.16   Let Σ be the covariance matrix of a random column vector $\mathbf{X}$. Write $\mathbf{Y} = \mathbf{X} - \boldsymbol{\mu}$, where $\boldsymbol{\mu}$ is the expectation vector of $\mathbf{X}$. Hence, $\Sigma = \mathbb{E}[\mathbf{Y}\mathbf{Y}^T]$. Show that Σ is positive semidefinite. That is, for any vector $\mathbf{u}$, we have $\mathbf{u}^T \Sigma \mathbf{u} \geqslant 0$.

1.17   Suppose $Y \sim \text{Gamma}(n, \lambda)$. Show that for all $x \geqslant 0$

$$\mathbb{P}(Y \leqslant x) = 1 - \sum_{k=0}^{n-1} \frac{e^{-\lambda x}(\lambda x)^k}{k!} . \tag{1.1}$$

1.18   Consider the random experiment where we draw uniformly and independently n numbers, $X_1, \ldots, X_n$, from the interval [0,1].

(a) Let M be the smallest of the n numbers. Express M in terms of $X_1, \ldots, X_n$.

(b) Determine the pdf of M.

1.19   Let $Y = e^X$, where $X \sim \text{N}(0, 1)$.

(a) Determine the pdf of Y.

(b) Determine the expected value of Y.

1.20   We select a point (X, Y) from the triangle $(0, 0) - (1, 0) - (1, 1)$ in such a way that X has a uniform distribution on $(0, 1)$ and the conditional distribution of Y given $X = x$ is uniform on $(0, x)$.

(a) Determine the joint pdf of X and Y.

(b) Determine the pdf of Y.

(c) Determine the conditional pdf of X given $Y = y$ for all $y \in (0, 1)$.

(d) Calculate $\mathbb{E}[X \,|\, Y = y]$ for all $y \in (0, 1)$.

(e) Determine the expectations of X and Y.

Poisson Processes

1.21   Let $\{N_t, t \geqslant 0\}$ be a Poisson process with rate $\lambda = 2$. Find

(a) $\mathbb{P}(N_2 = 1, N_3 = 4, N_5 = 5)$,

(b) $\mathbb{P}(N_4 = 3 \,|\, N_2 = 1, N_3 = 2)$,

(c) $\mathbb{E}[N_4 \,|\, N_2 = 2]$,

(d) $\mathbb{P}(N[2, 7] = 4, N[3, 8] = 6)$,

(e) $\mathbb{E}[N[4, 6] \,|\, N[1, 5] = 3]$.

1.22   Show that for any fixed $k \in \mathbb{N}$, $t > 0$, and $\lambda > 0$,

$$\lim_{n \to \infty} \binom{n}{k} \left(\frac{\lambda t}{n} \right)^k \left(1 - \frac{\lambda t}{n} \right)^{n-k} = \frac{(\lambda t)^k}{k!} e^{-\lambda t} .$$

(Hint: write out the binomial coefficient and use the fact that $\lim_{n \to \infty} \left(1 - \frac{\lambda t}{n} \right)^n = e^{-\lambda t}$.)

1.23   Consider the Bernoulli approximation in **Section 1.11**. Let $U_1, U_2, \ldots$ denote the times of success for the Bernoulli process X.

(a) Verify that the "intersuccess" times $U_1, U_2 - U_1, \ldots$ are independent and have a geometric distribution with parameter $p = \lambda h$.

(b) For small h and $n = \lfloor t/h \rfloor$, show that the relationship $\mathbb{P}(A_1 > t) \approx \mathbb{P}(U_1 > n)$ leads in the limit, as $n \to \infty$, to

$$\mathbb{P}(A_1 > t) = e^{-\lambda t}.$$

1.24 If $\{N_t, t \geqslant 0\}$ is a Poisson process with rate λ, show that for $0 \leqslant u \leqslant t$ and $j = 0, 1, 2, \ldots, n$,

$$\mathbb{P}(N_u = j \mid N_t = n) = \binom{n}{j} \left(\frac{u}{t}\right)^j \left(1 - \frac{u}{t}\right)^{n-j},$$

that is, the conditional distribution of N_u given $N_t = n$ is binomial with parameters n and u/t.

Markov Processes

1.25 Determine the (discrete) pdf of each X_n, $n = 0, 1, 2, \ldots$ for the random walk in **Example 1.10**. (That is, the random walk on the integers, with transition graph in Figure 1.1 and starting at 0.)

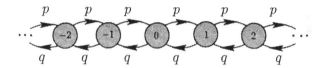

Figure 1.1 Transition graph for a random walk on $\mathbb{Z}$.

Also, calculate $\mathbb{E}[X_n]$ and the variance of X_n for each n.

1.26 Let $\{X_n, n \in \mathbb{N}\}$ be a Markov chain with state space $\{0, 1, 2\}$, transition matrix

$$P = \begin{pmatrix} 0.3 & 0.1 & 0.6 \\ 0.4 & 0.4 & 0.2 \\ 0.1 & 0.7 & 0.2 \end{pmatrix},$$

and initial distribution $\pi = (0.2, 0.5, 0.3)$. Determine
(a) $\mathbb{P}(X_1 = 2)$,
(b) $\mathbb{P}(X_2 = 2)$,
(c) $\mathbb{P}(X_3 = 2 \mid X_0 = 0)$,
(d) $\mathbb{P}(X_0 = 1 \mid X_1 = 2)$,
(e) $\mathbb{P}(X_1 = 1, X_3 = 1)$.

1.27 Consider two dogs harboring a total number of m fleas. Spot initially has b fleas and Lassie has the remaining $m - b$. The fleas have agreed on the following immigration policy: at every time $n = 1, 2 \ldots$ a flea is selected at random from the total population and that flea will jump from one dog to the other. Describe the flea population on Spot as a Markov chain and find its stationary distribution.

1.28 Classify the states of the Markov chain with the following transition matrix:

$$P = \begin{pmatrix} 0.0 & 0.3 & 0.6 & 0.0 & 0.1 \\ 0.0 & 0.3 & 0.0 & 0.7 & 0.0 \\ 0.3 & 0.1 & 0.6 & 0.0 & 0.0 \\ 0.0 & 0.1 & 0.0 & 0.9 & 0.0 \\ 0.1 & 0.1 & 0.2 & 0.0 & 0.6 \end{pmatrix}.$$

1.29   Consider the following snakes-and-ladders game in Figure 1.2. Let N be the number of tosses required to reach the finish using a fair die. Calculate the expectation of N using a computer.

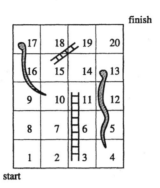

Figure 1.2   A snakes-and-ladders game.

1.30   Ms. Ella Brum walks back and forth between her home and her office every day. She owns three umbrellas, which are distributed over two umbrella stands (one at home and one at work). When it is not raining, Ms. Brum walks without umbrella. When it is raining, she takes one umbrella from the stand at the place of her departure, provided there is one available. Suppose the probability that it is raining at the time of any departure is p. Let X_n denote the number of umbrellas available at the place where Ella arrives after walk number n; $n = 1, 2, \ldots$, including the one that she possibly brings with her. Calculate the limiting probability that it rains and no umbrella is available.

1.31   A mouse is let loose in the maze of Figure 1.3. From each compartment the mouse chooses one of the adjacent compartments with equal probability, independent of the past. The mouse spends an exponentially distributed amount of time in each compartment. The mean time spent in each of the compartments 1, 3, and 4 is two seconds; the mean time spent in compartments 2, 5, and 6 is four seconds. Let $\{X_t, t \geqslant 0\}$ be the Markov jump process that describes the position of the mouse for times $t \geqslant 0$. Assume that the mouse starts in compartment 1 at time $t = 0$.

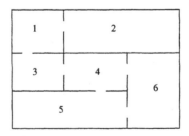

Figure 1.3   A maze.

What are the probabilities that the mouse will be found in compartments $1, 2, \ldots, 6$ at some time t far away in the future?

1.32 In an $M/M/\infty$-queueing system, customers arrive according to a Poisson process with rate a. Every customer who enters is immediately served by one of an infinite number of servers; hence, there is no queue. The service times are exponentially distributed, with mean $1/b$. All service and interarrival times are independent. Let X_t be the number of customers in the system at time t. Show that the limiting distribution of X_t, as $t \to \infty$, is Poisson with parameter a/b.

Optimization

1.33 Let $\mathbf{a}$ and $\mathbf{x}$ be n-dimensional column vectors. Show that $\nabla_{\mathbf{x}}\, \mathbf{a}^T \mathbf{x} = \mathbf{a}$.

1.34 Let A be a symmetric $n \times n$ matrix and let $\mathbf{x}$ be an n-dimensional column vector. Show that $\nabla_{\mathbf{x}}\, \frac{1}{2}\mathbf{x}^T A \mathbf{x} = A\mathbf{x}$. What is the gradient if A is not symmetric?

1.35 Show that the optimal distribution $\mathbf{p}^*$ in **Example 1.17** is given by the uniform distribution.

1.36 Derive the program **(1.78)**.

1.37 Consider the MinxEnt program

$$\min_{\mathbf{p}} \ \sum_{i=1}^{n} p_i \ln \frac{p_i}{q_i}$$

$$\text{subject to:} \quad \mathbf{p} \geqslant 0, \quad A\mathbf{p} = \mathbf{b}, \quad \sum_{i=1}^{n} p_i = 1 \,,$$

where $\mathbf{p}$ and $\mathbf{q}$ are probability distribution vectors and A is an $m \times n$ matrix.

  (a) Show that the Lagrangian for this problem is of the form

$$\mathcal{L}(\mathbf{p}, \lambda, \beta, \mu) = \mathbf{p}^T \xi(\mathbf{p}) - \lambda^T (A\mathbf{p} - \mathbf{b}) - \mu^T \mathbf{p} + \beta(\mathbf{1}^T \mathbf{p} - 1) \,.$$

  (b) Show that $p_i = q_i \exp(-\beta - 1 + \mu_i + \sum_{j=1}^{m} \lambda_j\, a_{ji})$ for $i = 1, \ldots, n$.
  (c) Explain why, as a result of the KKT conditions, the optimal μ^* must be equal to the zero vector.
  d) Show that the solution to this MinxEnt program is exactly the same as for the program where the nonnegativity constraints are omitted.

CHAPTER 2

RANDOM NUMBER, RANDOM VARIABLE, AND STOCHASTIC PROCESS GENERATION

2.1   Apply the inverse-transform method to generate a random variable from the discrete uniform distribution with pdf

$$f(x) = \begin{cases} \frac{1}{n+1}, & x = 0, 1, \ldots, n \\ 0, & \text{otherwise.} \end{cases}$$

2.2   Explain how to generate from the Beta$(1, \beta)$ distribution using the inverse-transform method.

2.3   Explain how to generate from the Weib(α, λ) distribution using the inverse-transform method.

2.4   Explain how to generate from the Pareto(α, λ) distribution using the inverse-transform method.

2.5   Many families of distributions are of *location-scale* type. That is, the cdf has the form

$$F(x) = F_0 \left(\frac{x - \mu}{\sigma} \right),$$

where μ is called the *location* parameter and σ the *scale* parameter, and F_0 is a fixed cdf that does not depend on μ and σ. The N(μ, σ^2) family of distributions is a good example,

Solutions Manual for SMCM, 2nd Edition. By D.P. Kroese, T. Taimre, Z.I. Botev, and R.Y. Rubinstein
Copyright © 2007 John Wiley & Sons, Inc.

where F_0 is the standard normal cdf. Write $F(x; \mu, \sigma)$ for $F(x)$. Let $X \sim F_0$ (that is, $X \sim F(x; 0, 1)$). Prove that $Y = \mu + \sigma X \sim F(x; \mu, \sigma)$. Thus, to sample from any cdf in a location-scale family, it suffices to know how to sample from F_0.

2.6 Apply the inverse-transform method to generate random variables from a *Laplace distribution* (that is, a shifted two-sided exponential distribution) with pdf

$$f(x) = \frac{\lambda}{2} e^{-\lambda|x-\theta|}, \quad -\infty < x < \infty, \quad (\lambda > 0).$$

2.7 Apply the inverse-transform method to generate a random variable from the *extreme value distribution*, which has cdf

$$F(x) = 1 - e^{-\exp(\frac{x-\mu}{\sigma})}, \quad -\infty < x < \infty, \quad (\sigma > 0).$$

2.8 Consider the triangular random variable with pdf

$$f(x) = \begin{cases} 0 & \text{if } x < 2a \text{ or } x \geqslant 2b \\[2mm] \dfrac{x - 2a}{(b-a)^2} & \text{if } 2a \leqslant x < a + b \\[2mm] \dfrac{(2b - x)}{(b-a)^2} & \text{if } a + b \leqslant x < 2b \end{cases}$$

(a) Derive the corresponding cdf F.

(b) Show that applying the inverse-transform method yields

$$X = \begin{cases} 2a + (b-a)\sqrt{2U} & \text{if } 0 \leqslant U < \frac{1}{2} \\[2mm] 2b + (a-b)\sqrt{2(1-U)} & \text{if } \frac{1}{2} \leqslant U < 1. \end{cases}$$

2.9 Present an inverse-transform algorithm for generating a random variable from the piecewise-constant pdf

$$f(x) = \begin{cases} C_i, & x_{i-1} \leqslant x \leqslant x_i, \ i = 1, 2, \ldots, n \\ 0 & \text{otherwise,} \end{cases}$$

where $C_i \geqslant 0$ and $x_0 < x_1 < \cdots < x_{n-1} < x_n$.

2.10 Let

$$f(x) = \begin{cases} C_i \, x, & x_{i-1} \leqslant x < x_i, \ i = 1, \ldots, n \\ 0, & \text{otherwise,} \end{cases}$$

where $C_i \geqslant 0$ and $x_0 < x_1 < \cdots < x_{n-1} < x_n$.

(a) Let $F_i = \sum_{j=1}^{i} \int_{x_{j-1}}^{x_j} C_j u \, du$, $i = 1, \ldots, n$. Show that the cdf F satisfies

$$F(x) = F_{i-1} + \frac{C_i}{2}\left(x^2 - x_{i-1}^2\right), \quad x_{i-1} \leqslant x < x_i, \ i = 1, \ldots, n.$$

(b) Describe an inverse-transform algorithm for random variable generation from $f(x)$.

2.11 A random variable is said to have a *Cauchy* distribution if its pdf is given by

$$f(x) = \frac{1}{\pi} \frac{1}{1+x^2}, \quad x \in \mathbb{R}.$$

Explain how one can generate Cauchy random variables using the inverse-transform method.

2.12 If X and Y are independent standard normal random variables, then $Z = X/Y$ has a Cauchy distribution. Show this. (Hint: first show that if U and $V > 0$ are continuous random variables with joint pdf $f_{U,V}$, then the pdf of $W = U/V$ is given by $f_W(w) = \int_0^\infty f_{U,V}(w\,v, v)\,v\,dv$.)

2.13 Verify the validity of the composition **Algorithm 2.3.4**.

2.14 Using the composition method, formulate and implement an algorithm for generating random variables from the following normal (Gaussian) mixture pdf:

$$f(x) = \sum_{i=1}^{3} p_i \frac{1}{b_i} \varphi\left(\frac{x - a_i}{b_i}\right),$$

where φ is the pdf of the standard normal distribution and $(p_1, p_2, p_3) = (1/2, 1/3, 1/6)$, $(a_1, a_2, a_3) = (-1, 0, 1)$, and $(b_1, b_2, b_3) = (1/4, 1, 1/2)$.

2.15 Verify that $C = \sqrt{2e/\pi}$ in Figure 2.1, where the pdf (times C) of the Exp(1) distribution and the positive standard normal pdf

$$f(x) = \sqrt{2/\pi}\,e^{-x^2/2}, \quad x \geqslant 0 \tag{2.1}$$

are plotted.

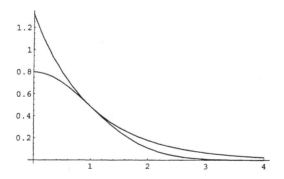

Figure 2.1 Bounding the positive normal density.

2.16 Prove that if $X \sim \text{Gamma}(\alpha, 1)$, then $X/\lambda \sim \text{Gamma}(\alpha, \lambda)$.

2.17 Let $X \sim \text{Gamma}(1 + \alpha, 1)$ and $U \sim \text{U}(0, 1)$ be independent. If $\alpha < 1$, then $XU^{1/\alpha} \sim \text{Gamma}(\alpha, 1)$. Prove this.

2.18 If $Y_1 \sim \text{Gamma}(\alpha, 1)$, $Y_2 \sim \text{Gamma}(\beta, 1)$, and Y_1 and Y_2 are independent, then

$$X = \frac{Y_1}{Y_1 + Y_2}$$

is Beta(α, β) distributed. Prove this.

2.19   Devise an acceptance-rejection algorithm for generating a random variable from the pdf f given in **(2.20)** (given in (2.1) above), using an $\text{Exp}(\lambda)$ proposal distribution. Which λ gives the largest acceptance probability?

2.20   The pdf of the truncated exponential distribution with parameter $\lambda = 1$ is given by

$$f(x) = \frac{e^{-x}}{1 - e^{-a}}, \quad 0 \leqslant x \leqslant a.$$

(a) Devise an algorithm for generating random variables from this distribution using the inverse-transform method.

(b) Construct a generation algorithm that uses the acceptance-rejection method with an $\text{Exp}(\lambda)$ proposal distribution.

(c) Find the efficiency of the acceptance-rejection method for the cases $a = 1$ and a approaching zero and infinity.

2.21   Let the random variable X have pdf

$$f(x) = \begin{cases} \frac{1}{4}, & 0 < x < 1 \\ x - \frac{3}{4}, & 1 \leqslant x \leqslant 2. \end{cases}$$

Generate a random variable from $f(x)$, using

(a) the inverse-transform method,

(b) the acceptance-rejection method, using the proposal density

$$g(x) = \frac{1}{2}, \quad 0 \leqslant x \leqslant 2.$$

2.22   Let the random variable X have pdf

$$f(x) = \begin{cases} \frac{1}{2}x, & 0 < x < 1 \\ \frac{1}{2}, & 1 \leqslant x \leqslant \frac{5}{2}. \end{cases}$$

Generate a random variable from $f(x)$, using

(a) the inverse-transform method

(b) the acceptance-rejection method, using the proposal density

$$g(x) = \frac{8}{25}x, \quad 0 \leqslant x \leqslant \frac{5}{2}.$$

2.23   Let X have a truncated geometric distribution, with pdf

$$f(x) = c\,p(1 - p)^{x-1}, \quad x = 1, \ldots, n,$$

where c is a normalization constant. Generate a random variable from $f(x)$, using

(a) the inverse-transform method,

(b) the acceptance-rejection method, with $G(p)$ as the proposal distribution. Find the efficiency of the acceptance-rejection method for $n = 2$ and $n = \infty$.

2.24   Generate a random variable $Y = \min_{i=1,\ldots,m} \max_{j=1,\ldots,r}\{X_{ij}\}$, assuming that the variables X_{ij}, $i = 1, \ldots, m$, $j = 1, \ldots, r$, are iid with common cdf $F(x)$, using the inverse-transform method. (Hint: Use the results for the distribution of order statistics in **Example 2.3**.)

2.25   Generate 100 Ber(0.2) random variables three times and produce bar graphs similar to those in **Figure 2.6**. Repeat for Ber(0.5).

2.26   Generate a homogeneous Poisson process with rate 100 on the interval $[0, 1]$. Use this to generate a nonhomogeneous Poisson process on the same interval with rate function

$$\lambda(t) = 100 \sin^2(10\,t), \quad t \geqslant 0 .$$

2.27   Generate and plot a realization of the points of a two-dimensional Poisson process with rate $\lambda = 2$ on the square $[0, 5] \times [0, 5]$. How many points fall in the square $[1, 3] \times [1, 3]$? How many do you expect to fall in this square?

2.28   Write a program that generates and displays 100 random vectors that are uniformly distributed within the ellipse

$$5\,x^2 + 21\,x\,y + 25\,y^2 = 9 .$$

2.29   Implement both random permutation algorithms in **Section 2.8**. Compare their performance.

2.30   Consider a random walk on the undirected graph in Figure 2.2. For example, if the random walk at some time is in state 5, it will jump to 3, 4, or 6 at the next transition, each with probability $1/3$.

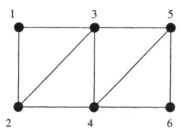

Figure 2.2 A graph.

(a) Find the one-step transition matrix for this Markov chain.
(b) Show that the stationary distribution is given by $\pi = (\frac{1}{9}, \frac{1}{6}, \frac{2}{9}, \frac{2}{9}, \frac{1}{6}, \frac{1}{9})$.
(c) Simulate the random walk on a computer and verify that, in the long run, the proportion of visits to the various nodes is in accordance with the stationary distribution.

2.31   Generate various sample paths for the random walk on the integers for $p = 1/2$ and $p = 2/3$.

2.32   Consider the $M/M/1$ queueing system of **Example 1.13**. Let X_t be the number of customers in the system at time t. Write a computer program to simulate the stochastic process $X = \{X_t\}$ by viewing X as a Markov jump process and applying **Algorithm 2.7.2**. Present sample paths of the process for the cases $\lambda = 1$, $\mu = 2$ and $\lambda = 10$, $\mu = 11$.

CHAPTER 3

SIMULATION OF DISCRETE-EVENT SYSTEMS

3.1   Consider the $M/M/1$ queueing system of **Example 1.13**. Let X_t be the number of customers in the system at time t. Write a computer program to simulate the stochastic process $X = \{X_t, t \geqslant 0\}$ using an event- or process-oriented DES approach. Present sample paths of the process for the cases $\lambda = 1$, $\mu = 2$ and $\lambda = 10$, $\mu = 11$.

3.2   Repeat the above simulation, but now assume $U(0, 2)$ interarrival times and $U(0, 1/2)$ service times (all independent).

3.3   Run the Matlab program of **Example 3.1** (or implement it in the computer language of your choice). This Matlab program is given below.

```
clear all;
T = 400;
x = 150; %initial amount of money.
xx = [150]; tt = [0];
t=0;
ev_list = inf*ones(3,2);        %record time, type
ev_list(1,:) = [7 + 3*rand, 1]; %schedule type 1 event
ev_list(2,:) = [25 + 10*rand,2]; %schedule type 2 event
ev_list(3,:) = [-log(rand),3];  %schedule type 3 event
ev_list = sortrows(ev_list,1);  % sort event list
while t < T
```

```
    t = ev_list(1,1);
    ev_type = ev_list(1,2);
    switch ev_type
        case 1
            x = x + 16*-log(rand);
            ev_list(1,:) = [7 + 3*rand + t, 1];
        case 2
            x = x + 100;
            ev_list(1,:) = [25 + 10*rand + t, 2];
        case 3
            x = x - (5 + randn);
            ev_list(1,:) = [-log(rand) + t, 3];
    end
    ev_list = sortrows(ev_list,1); % sort event list
    xx = [xx,x];
    tt = [tt,t];
end
plot(tt,xx)
```

Out of 1000 runs, how many lead to a negative account balance during the first 100 days? How does the process behave for large t?

3.4   Implement an event-oriented simulation program for the tandem queue. Let the interarrivals be exponentially distributed with mean 5, and let the service times be uniformly distributed on [3,6]. Plot realizations of the queue length processes of both queues.

3.5   Consider the repairman problem with two identical machines and one repairman. We assume that the lifetime of a machine has an exponential distribution with expectation 5 and that the repair time of a machine is exponential with expectation 1. All the lifetimes and repair times are independent of each other. Let X_t be the number of failed machines at time t.

(a) Verify that $X = \{X_t, t \geqslant 0\}$ is a birth-and-death process and give the corresponding birth and death rates.

(b) Write a program that simulates the process X according to **Algorithm 2.7.2** and use this to assess the fraction of time that both machines are out of order. Simulate from $t = 0$ to $t = 100,000$.

(c) Write an event-oriented simulation program for this process.

(d) Let the exponential life and repair times be uniformly distributed on $[0, 10]$ and $[0, 2]$, respectively (hence the expectations stay the same as before). Simulate from $t = 0$ to $t = 100,000$. How does the fraction of time that both machines are out of order change?

(e) Now simulate a repairman problem with the above life and repair times, but with five machines and three repairmen. Run again from $t = 0$ to $t = 100,000$.

3.6   Draw flow diagrams, such as in **Figure 3.10**, for all the processes in the tandem queue; see also **Figure 3.8**.

3.7   Consider the following queueing system. Customers arrive at a circle according to a Poisson process with rate λ. On the circle, which has circumference 1, a single server travels at constant speed α^{-1}. Upon arrival the customers choose their positions on the circle according to a uniform distribution. The server always moves toward the nearest customer, sometimes clockwise, sometimes counterclockwise. Upon reaching a customer, the server stops and serves the customer according to an exponential service time distribution with parameter μ. When the server is finished, the customer is removed from the circle and the server resumes his journey on the circle. Let $\eta = \lambda\alpha$, and let $X_t \in [0, 1]$ be the position of the server at time t. Furthermore, let N_t be the number of customers waiting on the circle at time t. Implement a simulation program for this so-called *continuous polling system with a "greedy" server*, and plot realizations of the processes $\{X_t, t \geqslant 0\}$ and $\{N_t, t \geqslant 0\}$, taking the parameters $\lambda = 1, \mu = 2$, for different values of α. Note that although the state space of $\{X_t, t \geqslant 0\}$ is continuous, the system is still a DEDS since between arrival and service events the system state changes deterministically.

3.8   Consider a *continuous flow line* consisting of three machines in tandem separated by two storage areas, or buffers, through which a continuous (fluid) stream of items flows from one machine to the next; see Figure 3.1.

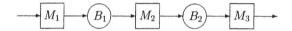

Figure 3.1   A flow line with three machines and two buffers (three-stage flow line).

Each machine $i = 1, 2, 3$ has a specific *machine speed* ν_i, which is the maximum rate at which it can transfer products from its upstream buffer to its downstream buffer. The lifetime of machine i has an exponential distribution with parameter λ_i. The repair of machine i starts immediately after failure and requires an exponential time with parameter μ_i. All life and repair times are assumed to be independent of each other. Failures are operation-independent. In particular, the failure rate of a "starved" machine (a machine that is idle because it does not receive input from its upstream buffer) is the same as that of a fully operational machine. The first machine has an unlimited supply.

Suppose all machine speeds are 1, the buffers are of equal size b, and all machines are identical with parameters $\lambda = 1$ and $\mu = 2$.

(a)   Implement an event- or process-oriented simulation program for this system.

(b)   Assess via simulation the average *throughput* of the system (the long-run amount of fluid that enters/leaves the system per unit of time) as a function of the buffer size b.

CHAPTER 4

STATISTICAL ANALYSIS OF DISCRETE-EVENT SYSTEMS

4.1   We wish to estimate $\ell = \int_{-2}^{2} e^{-x^2/2}\, dx = \int H(x)f(x)\, dx$ via Monte Carlo simulation using two different approaches: (1) defining $H(x) = 4\, e^{-x^2/2}$ and f the pdf of the $\mathsf{U}[-2, 2]$ distribution and (2) defining $H(x) = \sqrt{2\pi}\, I_{\{-2 \leqslant x \leqslant 2\}}$ and f the pdf of the $\mathsf{N}(0, 1)$ distribution.

   (a) For both cases estimate ℓ via the estimator $\widehat{\ell}$ in **(4.2)**, that is,

$$\widehat{\ell} = N^{-1} \sum_{i=1}^{N} H(\mathbf{X}_i) . \tag{4.1}$$

     Use a sample size of $N = 100$.

   (b) For both cases estimate the relative error κ of $\widehat{\ell}$ using $N = 100$.

   (c) Give a 95% confidence interval for ℓ for both cases using $N = 100$.

   (d) From part (b), assess how large N should be such that the relative width of the confidence interval is less than 0.001, and carry out the simulation with this N. Compare the result with the true value of ℓ.

Solutions Manual for SMCM, 2nd Edition. By D.P. Kroese, T. Taimre, Z.I. Botev, and R.Y. Rubinstein
Copyright © 2007 John Wiley & Sons, Inc.

4.2   Prove that the structure function of the bridge system given in Figure 4.1 is given by

$$H(\mathbf{x}) = 1 - (1 - x_1\,x_4)\,(1 - x_2\,x_5)\,(1 - x_1\,x_3\,x_5)\,(1 - x_2\,x_3\,x_4)\,.$$

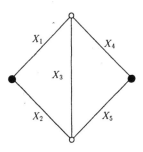

Figure 4.1   Bridge network.

4.3   Consider the bridge system in Figure 4.1. Suppose all link reliabilities are p. Show that the reliability of the system is $p^2(2 + 2\,p - 5\,p^2 + 2\,p^3)$.

4.4   Estimate the reliability of the bridge system in Figure 4.1 via **(4.2)** (which is (4.1) above) if the link reliabilities are $(p_1, \ldots, p_5) = (0.7,\ 0.6,\ 0.5,\ 0.4,\ 0.3)$. Choose a sample size such that the estimate has a relative error of about 0.01.

4.5   Consider the following sample performance:

$$H(\mathbf{X}) = \min\{X_1 + X_2,\ X_1 + X_4 + X_5,\ X_3 + X_4\}.$$

Assume that the random variables X_i, $i = 1, \ldots, 5$ are iid with common distribution

(a) **Gamma**(λ_i, β_i), where $\lambda_i = i$ and $\beta_i = i$.

(b) **Ber**(p_i), where $p_i = \dfrac{1}{2i}$.

Run a computer simulation with $N = 1000$ replications and find point estimates and 95% confidence intervals for $\ell = \mathbb{E}[H(\mathbf{X})]$.

4.6   Consider the precedence ordering of activities in Table 4.1. Suppose that durations of the activities (when actually started) are independent of each other, and all have exponential distributions with parameters 1.1, 2.3, 1.5, 2.9, 0.7, and 1.5, for activities $1, \ldots, 6$, respectively.

Table 4.1   Precedence ordering of activities.

Activity	1	2	3	4	5	6
Predecessor(s)	-	-	1	2,3	2,3	5

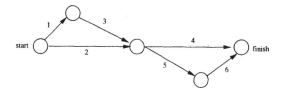

Figure 4.2 The PERT network corresponding to Table 4.1.

(a) Verify that the corresponding PERT graph is given by Figure 4.2.

(b) Identify the four possible paths from start to finish.

(c) Estimate the expected length of the critical path in (**4.5**), that is,

$$\ell = \mathbb{E}[H(\mathbf{X})] = \mathbb{E}\left[\max_{j=1,\ldots,p} \sum_{i \in \mathscr{P}_j} X_i\right],$$

where $\mathscr{P}_j$ is the j-th complete path from start to finish and p is the number of such paths, with a relative error of less than 5%.

4.7 Let $\{X_t, t = 0, 1, 2, \ldots\}$ be a random walk on the positive integers; see **Example 1.11**. Suppose that $p = 0.55$ and $q = 0.45$. Let $X_0 = 0$. Let Y be the maximum position reached after 100 transitions. Estimate the probability that $Y \geqslant 15$ and give a 95% confidence interval for this probability based on 1000 replications of Y.

4.8 Consider the $M/M/1$ queue. Let X_t be the number of customers in the system at time $t \geqslant 0$. Run a computer simulation of the process $\{X_t, t \geqslant 0\}$ with $\lambda = 1$ and $\mu = 2$, starting with an empty system. Let X denote the steady-state number of people in the system. Find point estimates and confidence intervals for $\ell = \mathbb{E}[X]$ using the batch means and regenerative methods as follows.

(a) For the batch means method run the system for a simulation time of 10,000, discard the observations in the interval $[0, 100]$, and use $N = 30$ batches.

(b) For the regenerative method, run the system for the same amount of simulation time (10,000) and take as regeneration points the times when an arriving customer finds the system empty.

(c) For both methods find the requisite simulation time that ensures a relative width of the confidence interval not exceeding 5%.

4.9 Let Z_n be the number of customers in an $M/M/1$ queueing system, as seen by the n-th arriving customer, $n = 1, 2, \ldots$. Suppose that the service rate is $\mu = 1$ and the arrival rate is $\lambda = 0.6$. Let Z be the steady-state queue length (as seen by an arriving customer far away in the future). Note that $Z_n = X_{T_n-}$, with X_t as in Problem 4.8, and T_n is the arrival epoch of the n-th customer; T_n- denotes the time just before T_n.

(a) Verify that $\ell = \mathbb{E}[Z] = 1.5$.

(b) Explain how to generate $\{Z_n, n = 1, 2, \ldots\}$ using a random walk on the positive integers, as in Problem 1.11.

(c) Find the point estimate of ℓ and a 95% confidence interval for ℓ using the batch means method. Use a sample size of 10^4 customers and $N = 30$ batches, throwing away the first $K = 100$ observations.

(d) Do the same as in (c) using the regenerative method instead.

(e) Assess the minimum length of the simulation run in order to obtain a 95% confidence interval with an absolute width w_a not exceeding 5%.

(f) Repeat (c), (d), and (e) with $\varrho = 0.8$ and discuss (c), (d), and (e) as $\varrho \to 1$.

4.10   Table 4.2 displays a realization of a Markov chain $\{X_t, t = 0, 1, 2, \ldots\}$ with state space $\{0, 1, 2, 3\}$ starting at 0. Let X be distributed according to the limiting distribution of this chain (assuming it has one).

Table 4.2 A realization of a Markov chain.

t	1	2	3	4	5	6	7	8	9	10	11	12	13	14	15
X_t	0	3	0	1	2	1	0	2	0	1	0	1	0	2	0

Find the point estimator, $\widehat{\ell}$, and the 95% confidence interval for $\ell = \mathbb{E}[X]$ using the regenerative method.

4.11   Let W_n be the *waiting time* of the n-th customer in a $GI/G/1$ queue, that is, the total time the customer spends waiting in the queue (thus excluding the service time). The waiting time process $\{W_n, n = 1, 2, \ldots\}$ follows the well-known *Lindley equation*:

$$W_{n+1} = \max\{W_n + S_n - A_{n+1}, 0\}, \quad n = 1, 2, \ldots,$$

where A_{n+1} is the interval between the n-th and $(n + 1)$-st arrivals, S_n is the service time of the n-th customer, and $W_1 = 0$ (the first customer does not have to wait and is served immediately).

(a) Explain why the Lindley equation holds.

(b) Find the point estimate and the 95% confidence interval for the expected waiting time for the fourth customer in an $M/M/1$ queue with $\varrho = 0.5$, $(\lambda = 1)$, starting with an empty system. Use $N = 5000$ replications.

(c) Find point estimates and confidence intervals for the expected average waiting time for customers $21, \ldots, 70$ in the same system as in b). Use $N = 5000$ replications. Hint: the point estimate and confidence interval required are for the following parameter:

$$\ell = \mathbb{E}\left[\frac{1}{50} \sum_{n=21}^{70} W_n \right].$$

4.12   Run a computer simulation of 1000 regenerative cycles of the (s, S) policy inventory model (see **Example 4.7**), where demands arrive according to a Poisson process with rate 2 (that is, $A \sim \text{Exp}(2)$) and the size of each demand follows a Poisson distribution with mean 2 (that is, $D \sim \text{Poi}(2)$). Take $s = 1$, $S = 6$, lead time $r = 2$, and initial value $X_0 = 4$. Find point estimates and confidence intervals for the quantity $\ell = \mathbb{P}(2 \leqslant X \leqslant 4)$, where X is the steady-state inventory position.

4.13   Simulate the two-state Markov chain $\{X_n\}$ in **Example 4.8**, starting in state 1, with transition matrix

$$P = \begin{pmatrix} p_{11} & p_{12} \\ p_{21} & p_{22} \end{pmatrix}$$

and cost matrix

$$C = (c_{ij}) = \begin{pmatrix} c_{11} & c_{12} \\ c_{21} & c_{22} \end{pmatrix} = \begin{pmatrix} 0 & 1 \\ 2 & 3 \end{pmatrix}.$$

Each transition from i to j incurs a cost of c_{ij}. Obtain a confidence interval for the long-run average cost using $p_{11} = 1/3$ and $p_{22} = 3/4$, with 1000 regeneration cycles.

4.14   Consider **Example 4.8** again (see Problem 4.13), with $p_{11} = 1/3$ and $p_{22} = 3/4$. Define $Y_i = (X_i, X_{i+1})$ and $H(Y_i) = c_{X_i, X_{i+1}}$, $i = 0, 1, \ldots$. Show that $\{Y_i\}$ is a regenerative process. Find the corresponding limiting/steady-state distribution and calculate $\ell = \mathbb{E}[H(Y)]$, where Y is distributed according to this limiting distribution. Check if ℓ is contained in the confidence interval obtained in Problem 4.13.

4.15   Consider the tandem queue of **Section 3.3.1**. Let X_t and Y_t denote the number of customers in the first and second queues at time t, including those who are possibly being served. Is $\{(X_t, Y_t), t \geqslant 0\}$ a regenerative process? If so, specify the regeneration times.

4.16   Consider the machine repair problem in Problem 3.5, with three machines and two repair facilities. Each repair facility can take only one failed machine. Suppose that the lifetimes are $\mathsf{Exp}(1/10)$ distributed and the repair times are $\mathsf{U}(0, 8)$ distributed. Let ℓ be the limiting probability that all machines are out of order.

(a) Estimate ℓ via the regenerative estimator $\widehat{\ell}$ in (**4.23**), that is,

$$\widehat{\ell} = \frac{\widehat{R}}{\widehat{\tau}},$$

where $\widehat{R} = N^{-1} \sum_{i=1}^{N} R_i$ and $\widehat{\tau} = N^{-1} \sum_{i=1}^{N} \tau_i$, using $N = 100$ regeneration cycles. Compute the 95% confidence interval (**4.26**) :

$$\left(\widehat{\ell} \pm \frac{z_{1-\alpha/2} \, S}{\widehat{\tau} \, N^{1/2}} \right),$$

where $S^2 = S_{11} - 2\widehat{\ell} S_{12} + \widehat{\ell}^2 S_{22}$ is the estimator of the asymptotic variance σ^2.

(b) Estimate the bias and mean square error of $\widehat{\ell}$ using the bootstrap method with a sample size of $B = 300$. (Hint: the original data is $\mathbf{X} = (X_1, \ldots, X_{100})$, where $X_i = (R_i, \tau_i)$, $i = 1, \ldots, 100$. Resample from these data using the empirical distribution.)

(c) Compute 95% bootstrap confidence intervals for ℓ using the normal and percentile methods with $B = 1000$ bootstrap samples.

CHAPTER 5

CONTROLLING THE VARIANCE

5.1 Consider the integral $\ell = \int_a^b H(x)\,dx = (b-a)\,\mathbb{E}[H(X)]$, with $X \sim \mathsf{U}(a,b)$. Let $X_1, \ldots, X_N$ be a random sample from $\mathsf{U}(a,b)$. Consider the estimators $\widehat{\ell} = \frac{1}{N}\sum_{i=1}^N H(X_i)$ and $\widehat{\ell}^{(a)} = \frac{1}{2N}\sum_{i=1}^N \{H(X_i) + H(b+a-X_i)\}$. Prove that if $H(x)$ is monotonic in x, then

$$\mathrm{Var}(\widehat{\ell}^{(a)}) \leqslant \frac{1}{2}\mathrm{Var}(\widehat{\ell}) .$$

In other words, using antithetic random variables is more accurate than using CMC.

5.2 Estimate the expected length of the shortest path for the bridge network in **Example 5.1**. That is, estimate $\ell = \mathbb{E}[H(\mathbf{X})]$, where

$$H(\mathbf{X}) = \min\{X_1 + X_4,\ X_1 + X_3 + X_5,\ X_2 + X_3 + X_4,\ X_2 + X_5\} .$$

Use both the CMC estimator **(5.8)**,

$$\widehat{\ell} = \frac{1}{N}\sum_{k=1}^N H(\mathbf{X}_k),$$

and the antithetic estimator **(5.9)**,

$$\widehat{\ell}^{(a)} = \frac{1}{N}\sum_{k=1}^{N/2}\left\{H(\mathbf{X}_k) + H(\mathbf{X}_k^{(a)})\right\},$$

where $\mathbf{X}_k = F^{-1}(\mathbf{U}_k)$ and $\mathbf{X}_k^{(a)} = F^{-1}(1 - \mathbf{U}_k)$. For both cases, take a sample size of $N = 100,000$. Assume that the lengths of the links $X_1, \ldots, X_5$ are exponentially distributed, with means $1, 1, 0.5, 2$, and 1.5, respectively. Compare the results.

5.3   Use the batch means method to estimate the expected stationary waiting time in a $GI/G/1$ queue via Lindley's equation for the case where the interarrival times are $\mathsf{Exp}(1/2)$ distributed and the service times are $\mathsf{U}[0.5, 2]$ distributed. Take a simulation run of $M = 10000$ customers, discarding the first $K = 100$ observations. Examine to what extent variance reduction can be achieved by using antithetic random variables.

5.4   Run the stochastic shortest path problem in **Example 5.4** and estimate the performance $\ell = \mathbb{E}[H(\mathbf{X})]$ from 1000 independent replications, using the given (C_1, C_2, C_3, C_4) as the vector of control variables and assuming that $X_i \sim \mathsf{Exp}(1)$, $i = 1, \ldots, 5$. Compare the results with those obtained with the CMC method.

5.5   Estimate the expected waiting time of the fourth customer in a $GI/G/1$ queue for the case where the interarrival times are $\mathsf{Exp}(1/2)$ distributed and the service times are $\mathsf{U}[0.5, 2]$ distributed. Use Lindley's equation and control variables, as described in **Example 5.5**. Generate $N = 1000$ replications of W_4 and provide a 95% confidence interval for $\mathbb{E}[W_4]$.

5.6   Prove that for any pair of random variables (U, V),

$$\mathrm{Var}(U) = \mathbb{E}[\,\mathrm{Var}(U \mid V)] + \mathrm{Var}(\,\mathbb{E}[U \mid V]\,)\,.$$

(Hint: Use the facts that $\mathbb{E}[U^2] = \mathbb{E}[\,\mathbb{E}[U^2 \mid V]\,]$ and $\mathrm{Var}(X) = \mathbb{E}[X^2] - (\mathbb{E}[X])^2$.)

5.7   Let $R \sim \mathsf{G}(p)$ and define $S_R = \sum_{i=1}^{R} X_i$, where $X_1, X_2, \ldots$ is a sequence of iid $\mathsf{Exp}(\lambda)$ random variables that are independent of R.

(a) Show that $S_R \sim \mathsf{Exp}(\lambda p)$. (Hint: the easiest way is to use transform methods and conditioning.)

(b) For $\lambda = 1$ and $p = 1/10$, estimate $\mathbb{P}(S_R > 10)$ using CMC with a sample size of $N = 1000$.

(c) Repeat (b), now using the conditional Monte Carlo estimator **(5.23)**:

$$\widehat{\ell}_c = \frac{1}{N} \sum_{k=1}^{N} F\left(x - \sum_{i=2}^{R_k} X_{ki}\right)\,.$$

Compare the results with those of (a) and (b).

5.8   Consider the random sum S_R in Problem 5.7, with parameters $p = 0.25$ and $\lambda = 1$. Estimate $\mathbb{P}(S_R > 10)$ via stratification using strata corresponding to the partition of events $\{R = 1\}, \{R = 2\}, \ldots, \{R = 7\}$, and $\{R > 7\}$. Allocate a total of $N = 10,000$ samples via both $N_i = p_i N$ and the optimal N_i^* in **(5.36)**,

$$N_i^* = N \frac{p_i \sigma_i}{\sum_{j=1}^{m} p_j \sigma_j}\,, \tag{5.1}$$

and compare the results. For the second method, use a simulation run of size 1000 to estimate the standard deviations $\{\sigma_i\}$.

5.9   Show that the solution to the minimization program

$$\min_{N_1, \ldots, N_m} \sum_{i=1}^{m} \frac{p_i^2 \sigma_i^2}{N_i} \quad \text{such that} \quad N_1 + \cdots + N_m = N\,,$$

is given by **(5.36)** (which is (5.1) above). This justifies stratified sampling **Theorem 5.5.1**.

5.10   Use **Algorithm 5.4.2** and **(5.27)** to estimate the reliability of the bridge reliability network in **Example 4.1** via permutation Monte Carlo. Consider two cases, where the link reliabilities are given by $\mathbf{p} = (0.3, 0.1, 0.8, 0.1, 0.2)$ and $\mathbf{p} = (0.95, 0.95, 0.95, 0.95, 0.95)$, respectively. Take a sample size of $N = 2000$.

5.11   Repeat Problem 5.10 using **Algorithm 5.4.3**. Compare the results.

5.12   This exercise discusses the counterpart of **Algorithm 5.4.3** involving minimal paths rather than minimal cuts. A state vector $\mathbf{x}$ in the reliability model of **Section 5.4.1** is called a *path vector* if $H(\mathbf{x}) = 1$. If in addition $H(\mathbf{y}) = 0$ for all $\mathbf{y} < \mathbf{x}$, then $\mathbf{x}$ is called the *minimal path vector*. The corresponding set $A = \{i : x_i = 1\}$ is called the *minimal path set*; that is, a minimal path set is a minimal set of components whose *functioning* ensures the functioning of the system. If $A_1, \ldots, A_m$ denote all the minimal paths sets, then the system is functioning if and only if all the components of at least one minimal path set are functioning.

   (a) Show that

$$H(\mathbf{x}) = \max_{k} \prod_{i \in A_k} x_i = 1 - \prod_{k=1}^{m} \left(1 - \prod_{i \in A_k} x_i \right). \qquad (5.2)$$

   (b) Define

$$Y_k = \prod_{i \in A_k} X_i, \quad k = 1, \ldots, m,$$

that is, Y_k is the indicator of the event that all components in A_i are functioning. Apply **Proposition 5.5.1** to the sum $S = \sum_{k=1}^{m} Y_k$ and devise an algorithm similar to **Algorithm 5.4.3** to estimate the reliability $r = \mathbb{P}(S > 0)$ of the system.

   (c) Test this algorithm on the bridge reliability network in **Example 4.1**.

5.13   Prove (see **(5.45)**) that the solution of

$$\min_{g} \text{Var}_g \left(H(\mathbf{X}) \frac{f(\mathbf{X})}{g(\mathbf{X})} \right)$$

is

$$g^*(\mathbf{x}) = \frac{|H(\mathbf{x})| f(\mathbf{x})}{\int |H(\mathbf{x})| f(\mathbf{x}) \, d\mathbf{x}}.$$

5.14   Let $Z \sim N(0, 1)$. Estimate $\mathbb{P}(Z > 4)$ via importance sampling, using the following shifted exponential sampling pdf:

$$g(x) = e^{-(x-4)}, \quad x \geqslant 4.$$

Choose N large enough to obtain accuracy to at least three significant digits and compare with the exact value.

5.15   Verify that the variance minimization program **(5.44)**,

$$\min_{g} \text{Var}_g \left(H(\mathbf{X}) \frac{f(\mathbf{X})}{g(\mathbf{X})} \right),$$

is equivalent to minimizing the Pearson χ^2 discrepancy measure (see **Remark 1.14.1**) between the zero-variance pdf g^* in **(5.46)** and the importance sampling density g. In this sense, the CE and VM methods are similar, because the CE method minimizes the Kullback–Leibler distance between g^* and g.

5.16   Repeat Problem 5.2 using importance sampling, where the lengths of the links are exponentially distributed with means $v_1, \ldots, v_5$. Write down the deterministic CE updating formulas and estimate these via a simulation run of size 1000 using $\mathbf{w} = \mathbf{u}$.

5.17   Consider the natural exponential family **(A.9)**,

$$f(\mathbf{x}; \boldsymbol{\theta}) = c(\boldsymbol{\theta})\, e^{\boldsymbol{\theta} \cdot \mathbf{t}(\mathbf{x})}\, h(\mathbf{x})\,, \tag{5.3}$$

where $\mathbf{t}(\mathbf{x}) = (t_1(\mathbf{x}), \ldots, t_m(\mathbf{x}))^T$ and $\boldsymbol{\theta} \cdot \mathbf{t}(\mathbf{x})$ is the inner product $\sum_{i=1}^{m} \theta_i t_i(\mathbf{x})$. Show that **(5.62)** (that is, $\mathbb{E}_\mathbf{u}\left[H(\mathbf{X})\,\nabla \ln f(\mathbf{X}; \mathbf{v})\right] = \mathbf{0}$, with $\mathbf{u} = \boldsymbol{\theta}_0$ and $\mathbf{v} = \boldsymbol{\theta}$) reduces to solving

$$\mathbb{E}_{\boldsymbol{\theta}_0}\left[H(\mathbf{X})\left(\frac{\nabla c(\boldsymbol{\theta})}{c(\boldsymbol{\theta})} + \mathbf{t}(\mathbf{X}) \right) \right] = \mathbf{0}\,. \tag{5.4}$$

5.18   As an application of (5.4), suppose that we wish to estimate the expectation of $H(X)$, with $X \sim \mathsf{Exp}(\lambda_0)$. Show that the corresponding CE optimal parameter is

$$\lambda^* = \frac{\mathbb{E}_{\lambda_0}[H(X)]}{\mathbb{E}_{\lambda_0}[H(X)X]}\,.$$

Compare with **(A.15)**:

$$v^* = \frac{\mathbb{E}_u[H(X)\,X]}{\mathbb{E}_u[H(X)]} = \frac{\mathbb{E}_w[H(X)\,W(X; u, w)\,X]}{\mathbb{E}_w[H(X)\,W(X; u, w)]}\,.$$

Explain how to estimate λ^* via simulation.

5.19   Let $X \sim \mathsf{Weib}(\alpha, \lambda_0)$. We wish to estimate $\ell = \mathbb{E}_{\lambda_0}[H(X)]$ via the SLR method, generating samples from $\mathsf{Weib}(\alpha, \lambda)$ — thus changing the scale parameter λ but keeping the scale parameter α fixed. Use (5.4) and Table 5.1 to show that the CE optimal choice for λ is

$$\lambda^* = \left(\frac{\mathbb{E}_{\lambda_0}[H(X)]}{\mathbb{E}_{\lambda_0}[H(X)\,X^\alpha]} \right)^{1/\alpha}\,.$$

Explain how we can estimate λ^* via simulation.

Table 5.1   The functions $c(\boldsymbol{\theta})$, $t_i(x)$ and $h(x)$ for commonly used distributions.

Distr.	$t_1(x),\ t_2(x)$	$c(\boldsymbol{\theta})$	$\theta_1,\ \theta_2$	$h(x)$
$\mathsf{Gamma}(\alpha, \lambda)$	$x,\ \ln x$	$\dfrac{(-\theta_1)^{\theta_2+1}}{\Gamma(\theta_2+1)}$	$-\lambda,\ \alpha - 1$	1
$\mathsf{N}(\mu, \sigma^2)$	$x,\ x^2$	$\dfrac{e^{\theta_1^2/(4\theta_2)}}{\sqrt{-\pi/\theta_2}}$	$\dfrac{\mu}{\sigma^2},\ -\dfrac{1}{2\sigma^2}$	1
$\mathsf{Weib}(\alpha, \lambda)$	$x^\alpha,\ \ln x$	$-\theta_1(\theta_2+1)$	$-\lambda^\alpha,\ \alpha - 1$	1
$\mathsf{Bin}(n, p)$	$x,\ -$	$(1 + e^{\theta_1})^{-n}$	$\ln\left(\dfrac{p}{1-p}\right),\ -$	$\dbinom{n}{x}$
$\mathsf{Poi}(\lambda)$	$x,\ -$	$e^{-e^{\theta_1}}$	$\ln \lambda,\ -$	$\dfrac{1}{x!}$
$\mathsf{G}(p)$	$x - 1,\ -$	$1 - e^{\theta_1}$	$\ln(1 - p),\ -$	1

5.20   Let $X_1, \ldots, X_n$ be independent $\mathsf{Exp}(1)$ distributed random variables. Let $\mathbf{X} = (X_1, \ldots, X_n)$ and $S(\mathbf{X}) = X_1 + \cdots + X_n$. We wish to estimate $\mathbb{P}(S(\mathbf{X}) \geqslant \gamma)$ via importance sampling, using $X_i \sim \mathsf{Exp}(\theta)$ for all i. Show that the CE optimal parameter θ^* is given by

$$\theta^* = \frac{\mathbb{E}[I_{\{S(\mathbf{X}) \geqslant \gamma\}}]}{\mathbb{E}[I_{\{S(\mathbf{X}) \geqslant \gamma\}} \, \overline{X}]} ,$$

with $\overline{X} = (X_1 + \cdots + X_n)/n$ and $\mathbb{E}$ indicating the expectation under the original distribution (where each $X_i \sim \mathsf{Exp}(1)$).

5.21   Consider Problem 5.19. Define $G(z) = z^{1/\alpha}/\lambda_0$ and $\widetilde{H}(z) = H(G(z))$.
(a) Show that if $Z \sim \mathsf{Exp}(1)$, then $G(Z) \sim \mathsf{Weib}(\alpha, \lambda_0)$.
(b) Explain how to estimate ℓ via the TLR method.
(c) Show that the CE optimal parameter for Z is given by

$$\theta^* = \frac{\mathbb{E}_\eta[\widetilde{H}(Z) \, W(Z; 1, \eta)]}{\mathbb{E}_\eta[\widetilde{H}(Z) \, Z \, W(Z; 1, \eta)]},$$

where $W(Z; 1, \eta)$ is the ratio of the $\mathsf{Exp}(1)$ and $\mathsf{Exp}(\eta)$ pdfs.

5.22   Assume that the expected performance can be written as $\ell = \sum_{i=1}^m a_i \ell_i$, where $\ell_i = \int H_i(\mathbf{x}) \, d\mathbf{x}$, and the a_i, $i = 1, \ldots, m$ are known coefficients. Let $Q(\mathbf{x}) = \sum_{i=1}^m a_i H_i(\mathbf{x})$. For any pdf g dominating $Q(\mathbf{x})$, the random variable

$$L = \sum_{i=1}^m a_i \frac{H_i(\mathbf{X})}{g(\mathbf{X})} = \frac{Q(\mathbf{X})}{g(\mathbf{X})} ,$$

where $\mathbf{X} \sim g$, is an unbiased estimator of ℓ — note that there is only one sample. Prove that L attains the smallest variance when $g = g^*$, with

$$g^*(\mathbf{x}) = \frac{|Q(\mathbf{x})|}{\int |Q(\mathbf{x})| \, d\mathbf{x}} ,$$

and that

$$\mathrm{Var}_{g^*}(L) = \left(\int |Q(\mathbf{x})| \, d\mathbf{x} \right)^2 - \ell^2 .$$

5.23   **The Hit-or-Miss Method.** Suppose that the sample performance function, H, is bounded on the interval $[0, b]$, say $0 \leqslant H(x) \leqslant c$ for $x \in [0, b]$. Let $\ell = \int H(x) \, dx = b \, \mathbb{E}[H(X)]$, with $X \sim \mathsf{U}[0, b]$. Define an estimator of ℓ by

$$\widehat{\ell}^h = \frac{bc}{N} \sum_{i=1}^N I_{\{Y_i < H(X_i)\}} ,$$

where $\{(X_i, Y_i) : j = 1, \ldots, N\}$ is a sequence of points uniformly distributed over the rectangle $[0, b] \times [0, c]$; see Figure 5.1 below. The estimator $\widehat{\ell}^h$ is called the *hit-or-miss estimator*, since a point (X, Y) is accepted or rejected depending on whether that point falls inside or outside the shaded area of Figure 5.1, respectively. Show that the hit-or-miss estimator has a larger variance than the CMC estimator,

$$\widehat{\ell} = \frac{b}{N} \sum_{i=1}^N H(X_i) ,$$

with $X_1, \ldots, X_N$ a random sample from $U[0, b]$.

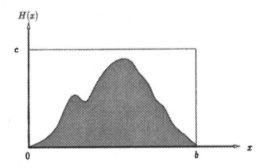

Figure 5.1   The hit-or-miss method.

CHAPTER 6

MARKOV CHAIN MONTE CARLO

6.1 Verify that the local balance equation (**6.3**),

$$\pi_i \, p_{ij} = \pi_j \, p_{ji}, \quad i, j \in \mathscr{X} \ ,$$

holds for the Metropolis–Hastings algorithm.

6.2 When running an MCMC algorithm, it is important to know when the transient (or *burn-in*) period has finished; otherwise, steady-state statistical analyses such as those in **Section 4.3.2** may not be applicable. In practice this is often done via a visual inspection of the sample path. As an example, run the random walk sampler with normal target distribution $N(10, 1)$ and proposal $Y \sim N(x, 0.01)$. Take a sample size of $N = 5000$. Determine roughly when the process reaches stationarity.

6.3 A useful tool for examining the behavior of a stationary process $\{X_t\}$ obtained, for example, from an MCMC simulation is the covariance function $R(t) = \text{Cov}(X_t, X_0)$; see **Example 6.4**. Estimate the covariance function for the process in Problem 6.2 and plot the results. In Matlab's *signal processing* toolbox this is implemented under the m-function xcov.m. Try different proposal distributions of the form $N(x, \sigma^2)$ and observe how the covariance function changes.

6.4   Implement the independence sampler with an $\mathsf{Exp}(1)$ target and an $\mathsf{Exp}(\lambda)$ proposal distribution for several values of λ. Similar to the importance sampling situation, things go awry when the sampling distribution gets too far from the target distribution, in this case when $\lambda > 2$. For each run use a sample size of 10^5 and start with $x = 1$.

(a) For each value $\lambda = 0.2, 1, 2$, and 5, plot a histogram of the data and compare it with the true pdf.

(b) For each value of the above values of λ, calculate the sample mean and repeat this for twenty independent runs. Make a dotplot of the data (plot them on a line) and notice the differences. Observe that for $\lambda = 5$ most of the sample means are below 1, and thus underestimate the true expectation 1, but a few are significantly greater. Observe also the behavior of the corresponding auto-covariance functions, both between the different λs and, for $\lambda = 5$, within the twenty runs.

6.5   Implement the random walk sampler with an $\mathsf{Exp}(1)$ target distribution, where Z (in the proposal $Y = x + Z$) has a double exponential distribution with parameter λ. Carry out a study similar to that in Problem 6.4 for different values of λ, say $\lambda = 0.1, 1, 5$, and 20. Observe that (in this case) the random walk sampler has more stable behavior than the independence sampler.

6.6   Let $\mathbf{X} = (X, Y)^T$ be a random column vector with a bivariate normal distribution with expectation vector $\mathbf{0} = (0, 0)^T$ and covariance matrix

$$\Sigma = \begin{pmatrix} 1 & \varrho \\ \varrho & 1 \end{pmatrix}.$$

(a) Show that $(Y \mid X = x) \sim \mathsf{N}(\varrho x, 1 - \varrho^2)$ and $(X \mid Y = y) \sim \mathsf{N}(\varrho y, 1 - \varrho^2)$.

(b) Write a systematic Gibbs sampler to draw 10^4 samples from the bivariate distribution $\mathsf{N}(\mathbf{0}, \Sigma)$ and plot the data for $\varrho = 0, 0.7$, and 0.9.

6.7   A remarkable feature of the Gibbs sampler is that the conditional distributions in **Algorithm 6.4.1** contain sufficient information to generate a sample from the joint one. The following result (due to Hammersley and Clifford) shows that it is possible to directly express the joint pdf in terms of the conditional ones. Namely,

$$f(x, y) = \frac{f_{Y \mid X}(y \mid x)}{\int \frac{f_{Y \mid X}(y \mid x)}{f_{X \mid Y}(x \mid y)} \, dy}.$$

Prove this. Generalize this to the n-dimensional case.

6.8   In the Ising model the *expected magnetization per spin* is given by

$$M(T) = \frac{1}{n^2} \mathbb{E}_{\pi_T} \left[\sum_i S_i \right],$$

where π_T is the Boltzmann distribution at temperature T. Estimate $M(T)$, for example via the Swendsen–Wang algorithm, for various values of $T \in [0, 5]$, and observe that the graph of $M(T)$ changes sharply around the critical temperature $T \approx 2.61$. Take $n = 20$ and use periodic boundaries.

6.9   Run Peter Young's Java applet at
        http://bartok.ucsc.edu/peter/java/ising/keep/ising.html
to gain a better understanding of how the Ising model works.

6.10   As in **Example 6.6**, let $\mathscr{X}^* = \{\mathbf{x} : \sum_{i=1}^{n} x_i = m, x_i \in \{0, \ldots, m\}, i = 1, \ldots, n\}$. Show that this set has $\binom{m+n-1}{n-1}$ elements.

6.11   In a simple model for a closed queueing network with n queues and m customers, it is assumed that the service times are independent and exponentially distributed, say with rate μ_i for queue i, $i = 1, \ldots, n$. After completing service at queue i, the customer moves to queue j with probability p_{ij}. The $\{p_{ij}\}$ are the so-called *routing probabilities*.

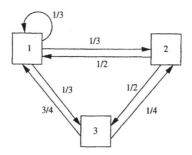

Figure 6.1   A closed queueing network.

It can be shown (see, for example, F. P. Kelly (1979) *Reversibility and Stochastic Networks* Wiley, Chichester) that the stationary distribution of the number of customers in the queues is of product form **(6.10)**, that is,

$$\pi(\mathbf{x}) = C \prod_{i=1}^{n} f_i(x_i), \quad \text{for } \sum_{i=1}^{n} x_i = m \,,$$

with f_i being the pdf of the $G(1 - y_i/\mu_i)$ distribution; thus, $f_i(x_i) \propto (y_i/\mu_i)^{x_i}$. Here the $\{y_i\}$ are constants that are obtained from the following set of *flow balance* equations:

$$y_i = \sum_j y_j \, p_{ji} \,, \quad i = 1, \ldots, n \,, \tag{6.1}$$

which has a one-dimensional solution space. Without loss of generality, y_1 can be set to 1 to obtain a unique solution.

Consider now the specific case of the network depicted in Figure 6.1, with $n = 3$ queues. Suppose the service rates are $\mu_1 = 2$, $\mu_2 = 1$, and $\mu_3 = 1$. The routing probabilities are given in the figure.

  (a) Show that a solution to (6.1) is $(y_1, y_2, y_3) = (1, 10/21, 4/7)$.

  (b) For $m = 50$ determine the exact normalization constant C.

  (c) Implement the procedure of **Example 6.6** to estimate C via MCMC and compare the estimate for $m = 50$ with the exact value.

6.12   Let $X_1, \ldots, X_n$ be a random sample from the $N(\mu, \sigma^2)$ distribution. Consider the following Bayesian model:

  • $f(\mu, \sigma^2) = 1/\sigma^2$;

  • $(\mathbf{x}_i \mid \mu, \sigma) \sim N(\mu, \sigma^2)$, $i = 1, \ldots, n$ independently.

Note that the prior for (μ, σ^2) is *improper*. That is, it is not a pdf in itself, but by obstinately applying Bayes' formula, it does yield a proper posterior pdf. In some sense it conveys the least amount of information about μ and σ^2. Let $\mathbf{x} = (x_1, \ldots, x_n)$ represent the data. The posterior pdf is given by

$$f(\mu, \sigma^2 \mid \mathbf{x}) = (2\pi\sigma^2)^{-n/2} \exp\left\{ -\frac{1}{2} \frac{\sum_i (x_i - \mu)^2}{\sigma^2} \right\} \frac{1}{\sigma^2}.$$

We wish to sample from this distribution via the Gibbs sampler.

(a) Show that $(\mu \mid \sigma^2, \mathbf{x}) \sim N(\bar{x}, \sigma^2/n)$, where $\bar{x}$ is the sample mean.

(b) Prove that

$$f(\sigma^2 \mid \mu, \mathbf{x}) \propto \frac{1}{(\sigma^2)^{n/2+1}} \exp\left(-\frac{n\, V_\mu}{2\,\sigma^2} \right), \tag{6.2}$$

where $V_\mu = \sum_i (x_i - \mu)^2 / n$ is the classical sample variance for known μ. In other words, $(1/\sigma^2 \mid \mu, \mathbf{x}) \sim \text{Gamma}(n/2, nV_\mu/2)$.

(c) Implement a Gibbs sampler to sample from the posterior distribution, taking $n = 100$. Run the sampler for 10^5 iterations. Plot the histograms of $f(\mu \mid \mathbf{x})$ and $f(\sigma^2 \mid \mathbf{x})$ and find the sample means of these posteriors. Compare them with the classical estimates.

(d) Show that the true posterior pdf of μ given the data is

$$f(\mu \mid \mathbf{x}) \propto \left((\mu - \bar{x})^2 + V \right)^{-n/2},$$

where $V = \sum_i (x_i - \bar{x})^2 / n$. (Hint: in order to evaluate the integral

$$f(\mu \mid \mathbf{x}) = \int_0^\infty f(\mu, \sigma^2 \mid \mathbf{x})\, d\sigma^2$$

write it first as $(2\pi)^{-n/2} \int_0^\infty t^{n/2-1} \exp(-\frac{1}{2} t\, c)\, dt$, where $c = n V_\mu$, by applying the change of variable $t = 1/\sigma^2$. Show that the latter integral is proportional to $c^{-n/2}$. Finally, apply the decomposition $V_\mu = (\bar{x} - \mu)^2 + V$.)

6.13 Suppose $f(\boldsymbol{\theta} \mid \mathbf{x})$ is the posterior pdf for some Bayesian estimation problem. For example, $\boldsymbol{\theta}$ could represent the parameters of a regression model based on the data $\mathbf{x}$. An important use for the posterior pdf is to make predictions about the distribution of other random variables. For example, suppose the pdf of some random variable Y depends on $\boldsymbol{\theta}$ via the conditional pdf $f(\mathbf{y} \mid \boldsymbol{\theta})$. The *predictive pdf* of Y given $\mathbf{x}$ is defined as

$$f(\mathbf{y} \mid \mathbf{x}) = \int f(\mathbf{y} \mid \boldsymbol{\theta}) f(\boldsymbol{\theta} \mid \mathbf{x})\, d\boldsymbol{\theta},$$

which can be viewed as the expectation of $f(\mathbf{y} \mid \boldsymbol{\theta})$ under the posterior pdf. Therefore, we can use Monte Carlo simulation to approximate $f(\mathbf{y} \mid \mathbf{x})$ as

$$f(\mathbf{y} \mid \mathbf{x}) \approx \frac{1}{N} \sum_{i=1}^N f(\mathbf{y} \mid \boldsymbol{\theta}_i),$$

where the sample $\{\boldsymbol{\theta}_i, i = 1, \ldots, N\}$ is obtained from $f(\boldsymbol{\theta} \mid \mathbf{x})$; for example, via MCMC.

As a concrete application, suppose that the independent measurement data $-0.4326, -1.6656, 0.1253, 0.2877, -1.1465$ come from some $N(\mu, \sigma^2)$ distribution. Define $\boldsymbol{\theta} = (\mu, \sigma^2)$. Let $Y \sim N(\mu, \sigma^2)$ be a new measurement. Estimate and draw the predictive pdf $f(y \mid \mathbf{x})$ from a sample $\boldsymbol{\theta}_1, \ldots, \boldsymbol{\theta}_N$ obtained via the Gibbs sampler of Problem 6.12. Take $N = 10,000$. Compare this with the "commonsense" Gaussian pdf with expectation $\bar{x}$ (sample mean) and variance s^2 (sample variance).

6.14 In the *zero-inflated Poisson* (ZIP) model, random data $X_1, \ldots, X_n$ are assumed to be of the form $X_i = R_i Y_i$, where the $\{Y_i\}$ have a Poi(λ) distribution and the $\{R_i\}$ have a Ber(p) distribution, all independent of each other. Given an outcome $\mathbf{x} = (x_1, \ldots, x_n)$, the objective is to estimate both λ and p. Consider the following hierarchical Bayes model:

- $p \sim \mathsf{U}(0,1)$ (prior for p),
- $(\lambda \mid p) \sim \mathsf{Gamma}(a, b)$ (prior for λ),
- $(r_i \mid p, \lambda) \sim \mathsf{Ber}(p)$ independently (from the model above),
- $(x_i \mid \mathbf{r}, \lambda, p) \sim \mathsf{Poi}(\lambda\, r_i)$ independently (from the model above),

where $\mathbf{r} = (r_1, \ldots, r_n)$ and a and b are known parameters. It follows that

$$f(\mathbf{x}, \mathbf{r}, \lambda, p) = \frac{b^a \lambda^{a-1} e^{-b\lambda}}{\Gamma(a)} \prod_{i=1}^n \frac{e^{-\lambda r_i} (\lambda r_i)^{x_i}}{x_i!} \, p^{r_i} (1-p)^{1-r_i} \; .$$

We wish to sample from the posterior pdf $f(\lambda, p, \mathbf{r} \mid \mathbf{x})$ using the Gibbs sampler.

(a) Show that

  1. $(\lambda \mid p, \mathbf{r}, \mathbf{x}) \sim \mathsf{Gamma}(a + \sum_i x_i, \; b + \sum_i r_i)$.
  2. $(p \mid \lambda, \mathbf{r}, \mathbf{x}) \sim \mathsf{Beta}(1 + \sum_i r_i, \; n + 1 - \sum_i r_i)$.
  3. $(r_i \mid \lambda, p, \mathbf{x}) \sim \mathsf{Ber}\left(\frac{p\, e^{-\lambda}}{p\, e^{-\lambda} + (1-p) I_{\{x_i = 0\}}} \right)$.

(b) Generate a random sample of size $n = 100$ for the ZIP model using parameters $p = 0.3$ and $\lambda = 2$.

(c) Implement the Gibbs sampler, generate a large (dependent) sample from the posterior distribution, and use this to construct 95% Bayesian confidence intervals for p and λ using the data in (b). Compare these with the true values.

6.15 * Show that μ in **(6.15)**, that is,

$$\mu(\mathbf{x}, \mathbf{y}) = f(\mathbf{x})\, q_{\mathbf{x}}(\mathbf{y}) \; ,$$

satisfies the local balance equations

$$\mu(\mathbf{x}, \mathbf{y})\, \mathbf{R}[(\mathbf{x}, \mathbf{y}), (\mathbf{x}', \mathbf{y}')] = \mu(\mathbf{x}', \mathbf{y}')\, \mathbf{R}[(\mathbf{x}', \mathbf{y}'), (\mathbf{x}, \mathbf{y})] \; .$$

Thus, μ is stationary with respect to $\mathbf{R}$, that is, $\mu \mathbf{R} = \mu$. Show that μ is also stationary with respect to $\mathbf{Q}$. Show, finally, that μ is stationary with respect to $\mathbf{P} = \mathbf{QR}$.

6.16 * This is to show that the systematic Gibbs sampler is a special case of the generalized Markov sampler. Take $\mathscr{Y}$ to be the set of indices $\{1, \ldots, n\}$ and define for the Q-step

$$\mathbf{Q}_{\mathbf{x}}(y, y') = \begin{cases} 1 & \text{if } y' = y + 1 \text{ or } y' = 1, y = n \\ 0 & \text{otherwise.} \end{cases}$$

Let the set of possible transitions $\mathscr{R}(\mathbf{x}, y)$ be the set of vectors $\{(\mathbf{x}', y)\}$ such that all coordinates of $\mathbf{x}'$ are the same as those of $\mathbf{x}$ except for possibly the y-th coordinate.

(a) Show that the stationary distribution of $\mathbf{Q}_{\mathbf{x}}$ is $q_{\mathbf{x}}(y) = 1/n$, for $y = 1, \ldots, n$.

(b) Show that

$$\mathbf{R}[(\mathbf{x}, y), (\mathbf{x}', y)] = \frac{f(\mathbf{x}')}{\displaystyle\sum_{(\mathbf{z}, y) \in \mathscr{R}(\mathbf{x}, y)} f(\mathbf{z})}, \quad \text{for } (\mathbf{x}', y) \in \mathscr{R}(\mathbf{x}, y) \; .$$

(c) Compare with **Algorithm 6.4.1**.

6.17 * Prove that the Metropolis–Hastings algorithm is a special case of the generalized Markov sampler. (Hint: let the auxiliary set $\mathscr{Y}$ be a copy of the target set $\mathscr{X}$, let $\mathbf{Q_x}$ correspond to the transition function of the Metropolis–Hastings algorithm (that is, $\mathbf{Q_x}(\cdot, \mathbf{y}) = q(\mathbf{x}, \mathbf{y})$), and define $\mathscr{R}(\mathbf{x}, \mathbf{y}) = \{(\mathbf{x}, \mathbf{y}), (\mathbf{y}, \mathbf{x})\}$. Use arguments similar to those for the Markov jump sampler (see **(6.20)**) to complete the proof.)

6.18 * Barker's and Hastings' MCMC algorithms differ from the symmetric Metropolis sampler only in that they define the acceptance ratio $\alpha(\mathbf{x}, \mathbf{y})$ to be respectively $f(\mathbf{y})/(f(\mathbf{x}) + f(\mathbf{y}))$ and $s(\mathbf{x}, \mathbf{y})/(1 + 1/\varrho(\mathbf{x}, \mathbf{y}))$, instead of $\min\{f(\mathbf{y})/f(\mathbf{x}), 1\}$. Here $\varrho(\mathbf{x}, \mathbf{y})$ is defined in **(6.6)**, that is,

$$\varrho(\mathbf{x}, \mathbf{y}) = \frac{f(\mathbf{y})\, q(\mathbf{y}, \mathbf{x})}{f(\mathbf{x})\, q(\mathbf{x}, \mathbf{y})} ,$$

and s is any symmetric function such that $0 \leqslant \alpha(\mathbf{x}, \mathbf{y}) \leqslant 1$. Show that both are special cases of the generalized Markov sampler. (Hint: take $\mathscr{Y} = \mathscr{X}$.)

6.19   Implement the simulated annealing algorithm for the n-queens problem suggested in **Example 6.13**. How many solutions can you find?

6.20   Implement the Metropolis–Hastings–based simulated annealing algorithm for the TSP in **Example 6.21**. Run the algorithm on some test problems from

> http://www.iwr.uni-heidelberg.de/groups/comopt/software/TSPLIB95/

6.21   Write a simulated annealing algorithm based on the random walk sampler to maximize the function

$$S(x) = \left| \frac{\sin^8(10x) + \cos^5(5x + 1)}{x^2 - x + 1} \right|, \quad x \in \mathbb{R} .$$

Use a $\mathrm{N}(x, \sigma^2)$ proposal function, given the current state x. Start with $x = 0$. Plot the current best function value against the number of evaluations of S for various values of σ and various annealing schedules. Repeat the experiments several times to assess what works best.

CHAPTER 7

SENSITIVITY ANALYSIS AND MONTE CARLO OPTIMIZATION

7.1   Consider the unconstrained minimization program

$$\min_u \ell(u) = \min_u \left\{ \mathbb{E}_u[X] + \frac{b}{u} \right\}, \quad u \in (0,1) , \tag{7.1}$$

where $X \sim \mathrm{Ber}(u)$.

(a) Show that the stochastic counterpart of $\nabla \ell(u) = 0$ can be written (see (**7.18**)) as

$$\widehat{\nabla \ell}(u) = \nabla \widehat{\ell}(u; v) = \frac{1}{v} \frac{1}{N} \sum_{i=1}^{N} X_i - \frac{b}{u^2} = 0 , \tag{7.2}$$

where $X_1, \ldots, X_N$ is a random sample from $\mathrm{Ber}(v)$.

(b) Assume that the sample $\{0,1,0,0,1,0,0,1,1,1,1,0,1,0,1,1,0,1,0,1,1\}$ was generated from $\mathrm{Ber}(v = 1/2)$. Show that the optimal solution u^* is estimated as

$$\widehat{u^*} = \left(\frac{b}{1.1} \right)^{1/2} .$$

Solutions Manual for SMCM, 2nd Edition. By D.P. Kroese, T. Taimre, Z.I. Botev, and R.Y. Rubinstein
Copyright © 2007 John Wiley & Sons, Inc.

7.2   Consider the unconstrained minimization program

$$\min_u \ell(u) = \min_u \{\mathbb{E}_u[X] + bu\}, \ u \in (0.5, 2.0) \ , \tag{7.3}$$

where $X \sim \mathsf{Exp}(u)$. Show that the stochastic counterpart of $\nabla \ell(u) = -\frac{1}{u^2} + b = 0$ can be written (see **(7.20)**) as

$$\nabla \widehat{\ell}(u; v) = \frac{1}{N} \sum_{i=1}^N X_i \frac{e^{-uX_i}(1 - uX_i)}{v\,e^{-vX_i}} + b = 0 \ , \tag{7.4}$$

where $X_1, \ldots, X_n$ is a random sample from $\mathsf{Exp}(v)$.

7.3   Prove equation **(7.25)**:

$$\mathbb{E}_v[W^2] = \mathbb{E}_u[W] = \prod_{k=1}^n \frac{1}{1 - \delta_k^2} \ ,$$

where $\delta_k = (u_k - v_k)/u_k, \ k = 1, \ldots, n$.

7.4   Show that $\nabla^k W(\mathbf{x}; \mathbf{u}, \mathbf{v}) = \mathcal{S}^{(k)}(\mathbf{u}; \mathbf{x})\, W(\mathbf{x}; \mathbf{u}, \mathbf{v})$ and hence prove **(7.16)**:

$$\widehat{\nabla^k \ell}(\mathbf{u}; \mathbf{v}) = \nabla^k \widehat{\ell}(\mathbf{u}; \mathbf{v}) = \frac{1}{N} \sum_{i=1}^N H(\mathbf{X}_i)\, \mathcal{S}^{(k)}(\mathbf{u}; \mathbf{X}_i)\, W(\mathbf{X}_i; \mathbf{u}, \mathbf{v}) \ .$$

7.5   Let $X_i \sim \mathsf{N}(u_i, \sigma_i^2), i = 1, \ldots, n$ be independent random variables. Suppose we are interested in sensitivities with respect to $\mathbf{u} = (u_1, \ldots, u_n)$ only. Show that for $i = 1, \ldots, n$,

$$\mathbb{E}_v[W^2] = \exp\left(\sum_{i=1}^n \frac{(u_i - v_i)^2}{\sigma_i^2}\right)$$

and

$$[\mathcal{S}^{(1)}(\mathbf{u}; \mathbf{X})]_i = \sigma_i^{-2}(x_i - u_i) \ .$$

7.6   Let the components $X_i, \ i = 1, \ldots, n$, of a random vector $\mathbf{X}$ be independent and distributed according to the exponential family

$$f_i(x_i; \mathbf{u}_i) = c_i(\mathbf{u}_i)\, e^{b_i(\mathbf{u}_i)t_i(x_i)}\, h_i(x_i) \ ,$$

where $b_i(\mathbf{u}_i)$, $t_i(x_i)$, and $h_i(x_i)$ are real-valued functions and $c_i(\mathbf{u}_i)$ is normalization constant. The corresponding pdf of $\mathbf{X}$ is given by

$$f(\mathbf{x}; \mathbf{u}) = c(\mathbf{u}) \exp\left(\sum_{i=1}^n b_i(\mathbf{u}_i)t_i(x_i)\right) h(\mathbf{x}) \ ,$$

where $\mathbf{u} = (\mathbf{u}_1^T, \ldots, \mathbf{u}_n^T)$, $c(\mathbf{u}) = \prod_{i=1}^n c_i(\mathbf{u}_i)$, and $h(\mathbf{x}) = \prod_{i=1}^n h_i(x_i)$.

(a) Show that $\mathrm{Var}_v(HW) = \frac{c(\mathbf{u})^2}{c(\mathbf{v})\,c(\mathbf{w})}\, \mathbb{E}_\mathbf{w}[H^2] - \ell(\mathbf{u})^2$, where $\mathbf{w}$ is determined by $b_i(\mathbf{w}_i) = 2b_i(\mathbf{u}_i) - b_i(\mathbf{v}_i), \ i = 1, \ldots, n$.

(b) Show that

$$\mathbb{E}_v[H^2 W^2] = \mathbb{E}_v[W^2]\, \mathbb{E}_\mathbf{w}[H^2] \ .$$

7.7   Consider the exponential pdf $f(x; u) = u \exp(-ux)$. Show that if $H(x)$ is a monotonically increasing function, then the expected performance $\ell(u) = \mathbb{E}_u[H(X)]$ is a monotonically decreasing convex function of $u \in (0, \infty)$.

7.8   Let $X \sim N(u, \sigma^2)$. Suppose that σ is known and fixed. For a given u, consider the function

$$\mathcal{L}(v) = \mathbb{E}_v[H^2 W^2].$$

(a) Show that if $\mathbb{E}_u[H^2] < \infty$ for all $u \in \mathbb{R}$, then $\mathcal{L}(v)$ is convex and continuous on $\mathbb{R}$. Show further that if, additionally, $\mathbb{E}_{u_n}[H^2] > 0$ for any u, then $\mathcal{L}(v)$ has a unique minimizer, v^*, over $\mathbb{R}$.

(b) Show that if $H^2(x)$ is monotonically increasing on $\mathbb{R}$, then $v^* > u$.

7.9   Let $X \sim N(u, \sigma^2)$. Suppose that u is known and consider the parameter σ. Note that the resulting exponential family is not of canonical form (that is, of the form (5.3)). However, parameterizing it by $\theta = \sigma^{-2}$ transforms it into canonical form, with $t(x) = -(x - u)^2/2$ and $c(\theta) = (2\pi)^{-1/2}\theta^{1/2}$.

(a) Show that

$$\mathbb{E}_\eta[W^2] = \frac{\theta}{\eta^{1/2}(2\theta - \eta)^{1/2}},$$

provided that $0 < \eta < 2\theta$.

(b) Show that, for a given θ, the function

$$\mathcal{L}(\eta) = \mathbb{E}_\eta[H^2 W^2]$$

has a unique minimizer, η^*, on the interval $(0, 2\theta)$, provided that the expectation $\mathbb{E}_\eta[H^2]$ is finite for all $\eta \in (0, 2\theta)$ and does not tend to 0 as η approaches 0 or 2θ. (Notice that this implies that the corresponding optimal value, $\sigma^* = \eta^{*-1/2}$, of the reference parameter, σ, is also unique.)

(c) Show that if $H^2(x)$ is strictly convex on $\mathbb{R}$, then $\eta^* < \theta$. (Notice that this implies that $\sigma^* > \sigma$.)

7.10   Consider the performance

$$H(X_1, X_2; u_3, u_4) = \min\{\max(X_1, u_3), \max(X_2, u_4)\},$$

where X_1 and X_2 have continuous densities $f(x_1; u_1)$ and $f(x_2; u_2)$, respectively. If we let $Y_1 = \max(X_1, u_3)$ and $Y_2 = \max(X_2, u_4)$ and write the performance as $\min(Y_1, Y_2)$, then Y_1 and Y_2 would take values u_3 and u_4 with nonzero probability. Hence, the random vector $\mathbf{Y} = (Y_1, Y_2)$ would not have a density function at point (u_3, u_4), since its distribution is a mixture of continuous and discrete ones. Consequently, the push-out method would fail in its current form. To overcome this difficulty, we carry out a transformation. We first write Y_1 and Y_2 as

$$Y_1 = u_3 \max\left(\frac{X_1}{u_3}, 1\right), \quad Y_2 = u_4 \max\left(\frac{X_2}{u_4}, 1\right)$$

and then replace $\mathbf{X} = (X_1, X_2)$ by the random vector $\widetilde{\mathbf{X}} = (\widetilde{X}_1, \widetilde{X}_2)$, where

$$\widetilde{X}_1 = \max\left(\frac{X_1}{u_3}, 1\right) \text{ and } \widetilde{X}_2 = \max\left(\frac{X_2}{u_4}, 1\right).$$

Prove that the density of the random vector $(\widetilde{X}_1, \widetilde{X}_2)$ is differentiable with respect to the variables (u_3, u_4), provided that both $\widetilde{X}_1$ and $\widetilde{X}_2$ are greater than 1.

7.11   **Delta method.** Let $X = (X_1, \ldots, X_n)$ and $Y = (Y_1, \ldots, Y_m)$ be random (column) vectors, with $Y = g(X)$ for some mapping g from $\mathbb{R}^n$ to $\mathbb{R}^m$. Let Σ_X and Σ_Y denote the corresponding covariance matrices. Suppose that X is close to its mean μ. A first-order Taylor expansion of g around μ gives

$$Y \approx g(\mu) + J_\mu(g)(X - \mu) \, ,$$

where $J_\mu(g)$ is the matrix of Jacobi of g (the matrix whose (i, j)-th entry is the partial derivative $\partial g_i / \partial x_j$) evaluated at μ. Show that, as a consequence,

$$\Sigma_Y \approx J_\mu(g) \Sigma_X J_\mu(g)^T \, .$$

This is called the *delta method* in statistics.

CHAPTER 8

THE CROSS-ENTROPY METHOD

8.1   In **Example 8.2**, show that the true CE-optimal parameter for estimating $\mathbb{P}(X \geqslant 32)$ is given by $v^* = 33$.

8.2   Write a CE program to reproduce **Table 8.1** in **Example 8.2**. Use the final reference parameter $\widehat{v}_3$ to estimate ℓ via importance sampling, using a sample size of $N_1 = 10^6$. Estimate the relative error and give an approximate 95% confidence interval. Check if the true value of ℓ is contained in this interval.

8.3   In **Example 8.2**, calculate the exact relative error for the importance sampling estimator $\widehat{\ell}$ when using the CE-optimal parameter $v^* = 33$ and compare it with the one estimated in Problem 8.2. How many samples are required to estimate ℓ with the same relative error using CMC?

8.4   Implement CE **Algorithm 8.2.1** for the stochastic shortest path problem in **Example 8.5** and reproduce **Table 8.3**.

8.5   Slightly modify the program used in Problem 8.4 to allow Weibull-distributed lengths. Reproduce **Table 8.4** and make a new table for $\alpha = 5$ and $\gamma = 2$ (the other parameters remain the same).

Solutions Manual for SMCM, 2nd Edition. By D.P. Kroese, T. Taimre, Z.I. Botev, and R.Y. Rubinstein
Copyright © 2007 John Wiley & Sons, Inc.   **41**

8.6 Make a table similar to **Table 8.4** by employing the standard CE method. That is, take $\text{Weib}(\alpha, v_i^{-1})$ as the importance sampling distribution for the i-th component and update the $\{v_i\}$ via **(8.6)**:

$$\max_{\mathbf{v}} \widehat{D}(\mathbf{v}) = \max_{\mathbf{v}} \frac{1}{N} \sum_{k=1}^{N} I_{\{S(\mathbf{X}_k) \geqslant \widehat{\gamma}_t\}} \ln f(\mathbf{X}_k; \mathbf{v}) .$$

8.7 Consider again the stochastic shortest path problem in **Example 8.5**, but now with nominal parameter $\mathbf{u} = (0.25, 0.5, 0.1, 0.3, 0.2)$. Implement the root-finding **Algorithm 8.2.3** to estimate for which level γ the probability ℓ is equal to 10^{-5}. Also give a 95% confidence interval for γ, for example using the bootstrap method.

8.8 Adapt the cost matrix in the max-cut program of **Table 8.10** and apply it to the dodecahedron max-cut problem in **Example 8.9**. Produce various optimal solutions and find out how many of these exist in total, disregarding the fivefold symmetry.

8.9 Consider the following symmetric cost matrix for the max-cut problem:

$$C = \begin{pmatrix} Z_{11} & B_{12} \\ B_{21} & Z_{22} \end{pmatrix}, \tag{8.1}$$

where Z_{11} is an $m \times m$ $(m < n)$ symmetric matrix in which all the upper-diagonal elements are generated from a $U(a, b)$ distribution (and all the lower-diagonal elements follow by symmetry), Z_{22} is an $(n - m) \times (n - m)$ symmetric matrix that is generated in the same way as Z_{11}, and all the other elements are c, apart from the diagonal elements, which are 0.

(a) Show that if $c > b(n - m)/m$, the optimal cut is given by $V^* = \{\{1, \ldots, m\}, \{m + 1, \ldots, n\}\}$.

(b) Show that the optimal value of the cut is $\gamma^* = cm(n - m)$.

(c) Implement and run the CE algorithm on this synthetic max-cut problem for a network with $n = 400$ nodes, with $m = 200$. Generate Z_{11} and Z_{22} from the $U(0, 1)$ distribution and take $c = 1$. For the CE parameters take $N = 1000$ and $\varrho = 0.1$. List for each iteration the best and worst of the elite samples and the Euclidean distance $\|\widehat{\mathbf{p}}_t - \mathbf{p}^*\| = \sqrt{(\widehat{p}_{t,i} - p_i^*)^2}$ as a measure of how close the reference vector is to the optimal reference vector $\mathbf{p}^* = (1, 1, \ldots, 1, 0, 0, \ldots, 0)$.

8.10 Consider a TSP with cost matrix $C = (c_{ij})$ defined by $c_{i,i+1} = 1$ for all $i = 1, 2, \ldots, n - 1$, and $c_{n,1} = 1$, while the remaining elements $c_{ij} \sim U(a, b)$, $j \neq i + 1$, $1 < a < b$, and $c_{ii} = 0$.

(a) Verify that the optimal permutation/tour is given by $\mathbf{x}^* = (1, 2, 3, \ldots, n)$, with minimal value $\gamma^* = n$.

(b) Implement a CE algorithm to solve an instance of this TSP for the case $n = 30$ and make a table of the performance, listing the best and worst of the elite samples at each iteration, as well as

$$p_t^{mm} = \min_{1 \leqslant i \leqslant n} \max_{1 \leqslant j \leqslant n} \widehat{p}_{t,ij} ,$$

$t = 1, 2, \ldots$, which corresponds to the min max value of the elements of the matrix $\widehat{P}_t$ at iteration t. Use $d = 3$, $\varrho = 0.01$, $N = 4500$, and $\alpha = 0.7$. Also, keep track of the overall best solution.

8.11 Run **Algorithm 8.3.1** on data from the URL

`http://www.iwr.uni-heidelberg.de/groups/comopt/software/TSPLIB95/atsp/`

and obtain a table similar to **Table 8.12**.

8.12   Select a TSP of your choice. Verify the following statements about the choice of CE parameters:

(a) By reducing ϱ or increasing α, the convergence is faster, but we can be trapped in a local minimum.

(b) By reducing ϱ, one needs to simultaneously decrease α, and vice versa, in order to avoid convergence to a local minimum.

(c) By increasing the sample size N, one can simultaneously reduce ϱ or (and) increase α.

8.13   Find out how many optimal solutions there are for the Hammersley TSP in **Example 8.11**.

8.14   Consider a complete graph with n nodes. With each edge from node i to j there is an associated cost c_{ij}. In the *longest path problem* the objective is to find the longest self-avoiding path from a certain *source* node to a *sink* node. Like the TSP this is a NP-complete problem.

(a) Assuming the source node is 1 and the sink node is n, formulate the longest path problem similar to the TSP. (The main difference is that the paths in the longest path problem can have different lengths.)

(b) Specify a path generation mechanism and the corresponding CE updating rules.

(c) Implement a CE algorithm for the longest path problem and apply it to a test problem.

8.15   Write a CE program that solves the eight-queens problem using the same configuration representation $\mathbf{X} = (X_1, \ldots, X_8)$ as in **Example 6.13**. A straightforward way to generate the configurations is to draw each X_i independently from a probability vector $(p_{i1}, \ldots, p_{i8})$, $i = 1, \ldots, 8$. Take $N = 500$, $\alpha = 0.7$, and $\varrho = 0.1$.

8.16   In the *permutation flow shop problem* n jobs have to be processed (in the same order) on m machines. The objective is to find the permutation of jobs that will minimize the *makespan*, that is, the time at which the last job is completed on machine m. Let $t(i, j)$ be the processing time for job i on machine j and let $\mathbf{x} = (x_1, x_2, \ldots, x_n)$ be a job permutation. Then the completion time $C(x_i, j)$ for job i on machine j can be calculated as follows.

$$
\begin{aligned}
C(x_1, 1) &= t(x_1, 1)\,, \\
C(x_i, 1) &= C(x_{i-1}, 1) + t(x_i, 1) \text{ for all } i = 2, \ldots, n\,, \\
C(x_1, j) &= C(x_1, j-1) + t(x_1, j) \text{ for all } j = 2, \ldots, m\,, \\
C(x_i, j) &= \max\{C(x_{i-1}, j), C(x_i, j-1)\} + t(x_i, j)
\end{aligned}
$$

$$\text{for all } i = 2, \ldots, n \text{ and } j = 2, \ldots, m\,.$$

The objective is to minimize $S(\mathbf{x}) = C(x_n, m)$. The trajectory generation for the permutation flow shop problem is similar to that of the TSP.

(a) Implement a CE algorithm to solve this problem.

(b) Run the algorithm for a benchmark problem from the Internet, for example `http://ina2.eivd.ch/Collaborateurs/etd/problemes.dir/ordonnancement.dir/ordonnancement.html` .

8.17   Verify the updating formulas **(8.46)** and **(8.47)**, that is,

$$\widehat{\mu}_{t,i} = \frac{\sum_{\mathbf{X}_k \in \mathscr{E}_t} X_{ki}}{|\mathscr{E}_t|}, \qquad i = 1, \dots, n,$$

and

$$\widehat{\sigma}_{t,i} = \sqrt{\frac{\sum_{\mathbf{X}_k \in \mathscr{E}_t} (X_{ki} - \widehat{\mu}_{t,i})^2}{|\mathscr{E}_t|}}, \qquad i = 1, \dots, n.$$

8.18   Plot Matlab's peaks function and verify that it has three local maxima.

8.19   Use the CE program in **Section A.5**:

```
n = 2;                                  % dimension
mu = [-3,-3]; sigma = 3*ones(1,n); N = 100; eps = 1E-5; rho = 0.1;
while max(sigma) > eps
  X = randn(N,n)*diag(sigma)+ mu(ones(N,1),:);
  SX= S(X);                             %Compute the  performance
  sortSX = sortrows([X, SX],n+1);
  Elite = sortSX((1-rho)*N:N,1:n); % elite samples
  mu = mean(Elite,1);                   % take sample  mean row-wise
  sigma = std(Elite,1);                 % take sample  st.dev. row-wise
  [S(mu),mu,max(sigma)]                 % output the  result
end

function out = S(X)
out =  3*(1-X(:,1)).^2.*exp(-(X(:,1).^2) - (X(:,2)+1).^2) ...
    - 10*(X(:,1)/5 - X(:,1).^3 - X(:,2).^5).*...
    exp(-X(:,1).^2-X(:,2).^2) - 1/3*exp(-(X(:,1)+1).^2 - X(:,2).^2);
end
```

to maximize the function $S(x) = e^{-(x-2)^2} + 0.8\,e^{-(x+2)^2}$. Examine the convergence of the algorithm by plotting the sequence of normal sampling densities in the same figure.

8.20   Use the CE method to minimize the *trigonometric function*

$$S(\mathbf{x}) = 1 + \sum_{i=1}^{n} 8\sin^2(\eta(x_i - x_i^*)^2) + 6\sin^2(2\eta(x_i - x_i^*)^2) + \mu(x_i - x_i^*)^2, \quad (8.2)$$

with $\eta = 7$, $\mu = 1$, and $x_i^* = 0.9, i = 1, \dots, n$. The global minimum $\gamma^* = 1$ is attained at $\mathbf{x}^* = (0.9, \dots, 0.9)$. Display the graph and density plot of this function and give a table for the evolution of the algorithm.

8.21   A well-known test case in continuous optimization is the *Rosenbrock* function (in n dimensions):

$$S(\mathbf{x}) = \sum_{i=1}^{n-1} 100\,(x_{i+1} - x_i^2)^2 + (x_i - 1)^2. \qquad (8.3)$$

The function has a global minimum $\gamma^* = 0$, attained at $\mathbf{x}^* = (1, 1, \dots, 1)$. Implement a CE algorithm to minimize this function for dimensions $n = 2, 5, 10$, and 20. Observe how injection (**Remark 8.7.1**) affects the accuracy and speed of the algorithm.

8.22   Suppose that $\mathscr{X}$ in **(8.15)**, that is,

$$\gamma^* = \max_{\mathbf{x} \in \mathscr{X}} S(\mathbf{x}) \,. \tag{8.4}$$

Here, $\mathscr{X}$ is a (possibly nonlinear) region defined by the following system of inequalities:

$$G_i(\mathbf{x}) \leqslant 0, \quad i = 1, \ldots, L \,. \tag{8.5}$$

The *proportional penalty* approach to constrained optimization is to modify the objective function as follows.

$$\widetilde{S}(\mathbf{x}) = S(\mathbf{x}) + \sum_{i=1}^{L} P_i(\mathbf{x}) \,, \tag{8.6}$$

where $P_i(\mathbf{x}) = C_i \max(G_i(\mathbf{x}), 0)$ and $C_i > 0$ measures the importance (cost) of the i-th penalty. It is clear that as soon as the constrained problem (8.4), (8.5) is reduced to the unconstrained one (8.4) — using (8.6) instead of S — we can again apply **Algorithm 8.3.1**.

Apply the proportional penalty approach to the constrained minimization of the Rosenbrock function of dimension 10 for the constraints below. List for each case the minimal value obtained by the CE algorithm (with injection, if necessary) and the CPU time. In all experiments, use $\varepsilon = 10^{-3}$ for the stopping criterion (stop if all standard deviations are less than ε) and $C = 1000$. Repeat the experiments ten times to check if indeed a global minimum is found.

(a) $\sum_{j=1}^{10} x_j \leqslant -8$

(b) $\sum_{j=1}^{10} x_j \geqslant 15$

(c) $\sum_{j=1}^{10} x_j \leqslant -8, \ \sum_{j=1}^{10} x_j^2 \geqslant 15$

(d) $\sum_{j=1}^{10} x_j \geqslant 15, \ \sum_{j=1}^{10} x_j^2 \leqslant 22.5$

8.23   Use the CE method to minimize the function

$$S(\mathbf{x}) = 1000 - x_1^2 - 2x_2^2 - x_3^2 - x_1 x_2 - x_1 x_3 \,,$$

subject to the constraints $x_j \geqslant 0, j = 1, 2, 3$, and

$$8 x_1 + 14 x_2 + 7 x_3 - 56 = 0 \,,$$
$$x_1^2 + x_2^2 + x_3^2 - 25 = 0 \,.$$

First, eliminate two of the variables by expressing x_2 and x_3 in terms of x_1. Note that this gives *two* different expressions for the pair (x_2, x_3). In the CE algorithm, generate the samples $\mathbf{X}$ by first drawing X_1 according to a truncated normal distribution on $[0,5]$. Then choose either the first or the second expression for (X_2, X_3) with equal probability. Verify that the optimal solution is approximately $\mathbf{x}^* = (3.51, 0.217, 3.55)$, with $S(\mathbf{x}^*) = 961.7$. Give the solution and the optimal value in seven significant digits.

8.24   Add $U(-0.1, 0.1)$, $N(0, 0.01)$, and $N(0, 1)$ noise to the objective function in Problem 8.19. Formulate an appropriate stopping criterion, for example based on $\widehat{\sigma}_t$. For each case observe how $\widehat{\gamma}_t$, $\widehat{\mu}_t$, and $\widehat{\sigma}_t$ behave.

8.25   Add $N(0,1)$ noise to the Matlab peaks function and apply the CE algorithm to find the global maximum. Display the contour plot and the path followed by the mean vectors $\{\widehat{\boldsymbol{\mu}}_t\}$, starting with $\widehat{\boldsymbol{\mu}}_0 = (1.3, -2.7)$ and using $N = 200$ and $\varrho = 0.1$. Stop when all standard deviations are less than $\varepsilon = 10^{-3}$. In a separate plot, display the evolution of the worst and best of the elite samples ($\widehat{\gamma}_t$ and S_t^*) at each iteration of the CE algorithm. In addition, evaluate and plot the noisy objective function in $\widehat{\boldsymbol{\mu}}_t$ for each iteration. Observe that, in contrast to the deterministic case, the $\{\widehat{\gamma}_t\}$ and $\{S_t^*\}$ do not converge to γ^* because of the noise, but eventually $S(\widehat{\boldsymbol{\mu}}_t)$ fluctuates around the optimum γ^*. More importantly, observe that the means $\{\widehat{\boldsymbol{\mu}}_t\}$ do converge to the optimal $\mathbf{x}^*$.

8.26   Select a particular instance (cost matrix) of the synthetic TSP in Problem 8.10. Make this TSP *noisy* by defining the random cost Y_{ij} from i to j in **(8.48)**, that is,

$$\widehat{S}(\mathbf{x}) = \sum_{i=1}^{n-1} Y_{x_i, x_{i+1}} + Y_{x_n, x_1} \, ,$$

to be $\mathrm{Exp}(c_{ij}^{-1})$ distributed. Apply the CE **Algorithm 8.3.1** to the noisy problem and compare the results with those in the deterministic case. Display the evolution of the algorithm in a graph, plotting the maximum distance, $\max_{i,j} |\widehat{p}_{t,ij} - p_{ij}^*|$, as a function of t.

CHAPTER 9

COUNTING VIA MONTE CARLO

9.1 Prove the upper bound **(9.21)**:

$$h(\varepsilon, p) = -\frac{1}{p} \ln \left(1 + \frac{\varepsilon p}{1 - p} \right) + (1 - \varepsilon)\theta^* \ .$$

9.2 Prove the upper bound **(9.22)**:

$$\ell_U \leqslant \exp \left(-\frac{np\varepsilon^2}{2 + \frac{2}{3}\varepsilon} \right) \ .$$

9.3 Consider **Problem 8.9**. Implement and run a PME algorithm on this synthetic max-cut problem for a network with $n = 400$ nodes and $m = 200$. Compare with the CE algorithm.

9.4 Let $\{A_i\}$ be an arbitrary collection of subsets of some finite set $\mathcal{X}$. Show that

$$|\cup_i A_i| = \sum |A_i| - \sum_{i<j} |A_i \cap A_j| + \sum_{i<j<k} |A_i \cap A_j \cap A_k| - \cdots$$

This is the useful *inclusion–exclusion* principle.

9.5   A famous problem in combinatorics is the *distinct representatives* problem, which is formulated as follows. Given a set $\mathscr{A}$ and subsets $\mathscr{A}_1,\ldots,\mathscr{A}_n$ of $\mathscr{A}$, is there a vector $\mathbf{x} = (x_1,\ldots,x_n)$ such that $x_i \in \mathscr{A}_i$ for each $i = 1,\ldots,n$ and the $\{x_i\}$ are all distinct (that is, $x_i \neq x_j$ if $i \neq j$)?

(a) Suppose, for example, that $\mathscr{A} = \{1,2,3,4,5\}$, $\mathscr{A}_1 = \{1,2,5\}$, $\mathscr{A}_2 = \{1,4\}$, $\mathscr{A}_3 = \{3,5\}$, $\mathscr{A}_4 = \{3,4\}$, and $\mathscr{A}_5 = \{1\}$. Count the total number of distinct representatives.

(b) Argue why the total number of distinct representatives in the above problem is equal to the *permanent* of the following matrix A.

$$A = \begin{pmatrix} 1 & 1 & 0 & 0 & 1 \\ 1 & 0 & 0 & 1 & 0 \\ 0 & 0 & 1 & 0 & 1 \\ 0 & 0 & 1 & 1 & 0 \\ 1 & 0 & 0 & 0 & 0 \end{pmatrix}.$$

9.6   Let $X_1,\ldots,X_n$ be independent random variables, each with marginal pdf f. Suppose we wish to estimate $\ell = \mathbb{P}_f(X_1 + \cdots + X_n \geqslant \gamma)$ using MinxEnt. For the prior pdf one could choose $h(\mathbf{x}) = f(x_1)f(x_2)\cdots f(x_n)$, that is, the joint pdf. We consider only a single constraint in the MinxEnt program, namely, $S(\mathbf{x}) = x_1 + \cdots + x_n$. As in **(9.34)**, the solution to this program is given by

$$g(\mathbf{x}) = c\,h(\mathbf{x})\,e^{\lambda S(\mathbf{x})} = c \prod_{j=1}^{n} e^{\lambda x_j} f(x_j) ,$$

where $c = 1/\mathbb{E}_h[e^{\lambda S(\mathbf{X})}] = (\mathbb{E}_f[e^{\lambda X}])^{-n}$ is a normalization constant and λ satisfies **(9.35)**:

$$\frac{\mathbb{E}_h\left[S(\mathbf{X})\exp\{\lambda\,S(\mathbf{X})\}\right]}{\mathbb{E}_h\left[\exp\{\lambda\,S(\mathbf{X})\}\right]} = \gamma .$$

Show that the new marginal pdfs are obtained from the old ones by an *exponential twist*, with twisting parameter $-\lambda$; see also **(A.13)**.

9.7   Problem 9.6 can be generalized to the case where $S(\mathbf{x})$ is a coordinatewise separable function, that is,

$$S(\mathbf{x}) = \sum_{i=1}^{n} S_i(x_i) ,$$

and the components $\{X_i\}$ are independent under the prior pdf $h(\mathbf{x})$. Show that also in this case the components under the optimal MinxEnt pdf $g(\mathbf{x})$ are independent and determine the marginal pdfs.

9.8   Let $\mathscr{X}$ be the set of permutations $\mathbf{x} = (x_1,\ldots,x_n)$ of the numbers $1,\ldots,n$, and let

$$S(\mathbf{x}) = \sum_{j=1}^{n} j\,x_j . \tag{9.1}$$

Let $\mathscr{X}^* = \{\mathbf{x} : S(\mathbf{x}) \geq \gamma\}$, where γ is chosen such that $|\mathscr{X}^*|$ is very small relative to $|\mathscr{X}| = n!$.

Implement a randomized algorithm to estimate $|\mathscr{X}^*|$ based on **(9.9)**, that is,

$$|\mathscr{X}^*| = |\mathscr{X}_0| \prod_{j=1}^m \frac{|\mathscr{X}_j|}{|\mathscr{X}_{j-1}|} \,,$$

using $\mathscr{X}_j = \{\mathbf{x} : S(\mathbf{x}) \geq \gamma_j\}$ for some sequence of $\{\gamma_j\}$ with $0 = \gamma_0 < \gamma_1 < \cdots < \gamma_r = \gamma$. Estimate each quantity $\mathbb{P}_U(\mathbf{X} \in \mathscr{X}_k \mid \mathbf{X} \in \mathscr{X}_{k-1})$ using the Metropolis–Hastings algorithm for drawing from the uniform distribution on $\mathscr{X}_{k-1}$. Define two permutations $\mathbf{x}$ and $\mathbf{y}$ as neighbors if one can be obtained from the other by swapping two indices, for example $(1,2,3,4,5)$ and $(2,1,3,4,5)$.

9.9   Write the Lagrangian dual problem for the MinxEnt problem with constraints in **Remark 9.5.1**:

$$\begin{cases} \min_g \mathcal{D}(g,h) = \min_g \int \ln \frac{g(\mathbf{x})}{h(\mathbf{x})} \, g(\mathbf{x}) \, \mathrm{d}\mathbf{x} = \min_g \mathbb{E}_g \left[\ln \frac{g(\mathbf{X})}{h(\mathbf{X})} \right] \\[2mm] \text{s.t.} \quad \int S_i(\mathbf{x}) g(\mathbf{x}) \, \mathrm{d}\mathbf{x} = \mathbb{E}_g[S_i(\mathbf{X})] = \gamma_i, \quad i = 1, \dots, m \,, \\[2mm] \mathbb{E}_g[S_i(\mathbf{X})] \geqslant \gamma_i, \quad i = m+1, \dots, m+M \\[2mm] \int g(\mathbf{x}) \, \mathrm{d}\mathbf{x} = 1 \,. \end{cases} \qquad (9.2)$$

Here g and h are n-dimensional pdfs, $S_i(\mathbf{x})$, $i = 1, \dots, m$, are given functions, and $\mathbf{x}$ is an n-dimensional vector.

CHAPTER 10

APPENDIX

10.1   Prove **(A.8)**:

$$\mathbb{P}\left(X_1 = x_1, \ldots, X_n = x_n \;\middle|\; \sum_{i=1}^{n} X_i = k\right) = \frac{\prod_{i=1}^{n} w_i^{x_i}}{c} \ .$$

10.2   Let X and Y be Gaussian random vectors, with joint distribution given by

$$\begin{pmatrix} X \\ Y \end{pmatrix} \sim \mathsf{N}\left(\underbrace{\begin{pmatrix} \mu_1 \\ \mu_2 \end{pmatrix}}_{\mu}, \underbrace{\begin{pmatrix} \Sigma_{11} & \Sigma_{12} \\ \Sigma_{21} & \Sigma_{22} \end{pmatrix}}_{\Sigma}\right).$$

(a)   Defining $S = \Sigma_{12}\Sigma_{22}^{-1}$, show that

$$\begin{pmatrix} I & -S \\ 0 & I \end{pmatrix} \Sigma \begin{pmatrix} I & 0 \\ -S^T & I \end{pmatrix} = \begin{pmatrix} \Sigma_{11} - S\Sigma_{21} & 0 \\ 0 & \Sigma_{22} \end{pmatrix}.$$

(b) Using the above result, show that for any vectors u and v

$$(u^T \ v^T)\Sigma^{-1} \begin{pmatrix} u \\ v \end{pmatrix} = (u^T - v^T S^T)\widetilde{\Sigma}^{-1}(u - Sv) + v^T \Sigma_{22}^{-1} v \, ,$$

where $\widetilde{\Sigma} = (\Sigma_{11} - S\Sigma_{21})$.

(c) The joint pdf of X and Y is given by

$$f(x, y) = c_1 \exp\left[-\frac{1}{2}(x^T - \mu_1^T \ \ y^T - \mu_2^T)\Sigma^{-1} \begin{pmatrix} x - \mu_1 \\ y - \mu_2 \end{pmatrix} \right],$$

for some constant c_1. Using b), show that the conditional pdf $f(x \,|\, y)$ is of the form

$$f(x \,|\, y) = c_2(y) \exp\left[-\frac{1}{2}(x^T - \tilde{\mu}^T)\,\widetilde{\Sigma}^{-1}\,(x - \tilde{\mu}) \right],$$

with $\tilde{\mu} = \mu_1 + S(y - \mu_2)$, and where $c_2(y)$ is some function of y (need not be specified). This proves that

$$(X \,|\, Y = y) \sim N\left(\mu_1 + \Sigma_{12}\Sigma_{22}^{-1}(y - \mu_2), \ \Sigma_{11} - \Sigma_{12}\Sigma_{22}^{-1}\Sigma_{12}^T \right) \, .$$

PART II

SOLUTIONS

CHAPTER 11

PRELIMINARIES

1.1 (a) First, $\mathbb{P}(A \cup A^c) = \mathbb{P}(\Omega) = 1$. Second, by the sum rule **(1.1)**, we have $\mathbb{P}(A \cup A^c) = \mathbb{P}(A) + \mathbb{P}(A^c)$. Combining gives $\mathbb{P}(A^c) = 1 - \mathbb{P}(A)$.

   (b) Write $A \cup B$ as the disjoint union of A and $B \cap A^c$. Then, again by the sum rule, $\mathbb{P}(A \cup B) = \mathbb{P}(A) + \mathbb{P}(B \cap A^c)$. Similarly, $B = (A \cap B) \cup (B \cap A^c)$, so that $\mathbb{P}(B) = \mathbb{P}(A \cap B) + \mathbb{P}(B \cap A^c)$. Combining these two equations gives $\mathbb{P}(A \cup B) = \mathbb{P}(A) + \mathbb{P}(B) - \mathbb{P}(A \cap B)$.

1.2 By applying the definition of conditional probability **(1.3)** to both $\mathbb{P}(B \mid A)$ and $\mathbb{P}(C \mid A \cap B)$, we find

$$\mathbb{P}(A) \, \mathbb{P}(B \mid A) \, \mathbb{P}(C \mid A \cap B) = \mathbb{P}(A) \frac{\mathbb{P}(A \cap B)}{\mathbb{P}(A)} \frac{\mathbb{P}(A \cap B \cap C)}{\mathbb{P}(A \cap B)} = \mathbb{P}(A \cap B \cap C) \,,$$

which proves the product rule **(1.4)** for the case of three events.

1.3 Let A_i be the event that the i-th draw yields a black ball, $i = 1, 2, 3$. By applying the product rule **(1.4)**, we find

$$\mathbb{P}(A_1 \cap A_2 \cap A_3) = \mathbb{P}(A_1) \, \mathbb{P}(A_2 \mid A_1) \, \mathbb{P}(A_3 \mid A_1 \cap A_2) = \frac{5}{10} \frac{4}{9} \frac{3}{8} = \frac{1}{12} \,.$$

1.4   Let A_i be the event that the i-th toss yields heads, $i = 1, 2, \ldots$. The $\{A_i\}$ are independent events, with $\mathbb{P}(A_i) = p$ and $\mathbb{P}(A_i^c) = 1 - p, i = 1, 2, \ldots$. Therefore, for every $k = 1, 2, \ldots$,

$$\mathbb{P}(X = k) = \mathbb{P}(A_1^c \cap A_2^c \cap \cdots \cap A_{k-1}^c \cap A_k) = \underbrace{(1 - p) \cdots (1 - p)}_{k-1 \text{ times}} p = (1 - p)^{k-1} p,$$

which means that $X \sim G(p)$.

1.5 (a) It follows from the product rule (1.4) that

$$\mathbb{P}(N > n) = \frac{365}{365} \times \frac{364}{365} \times \cdots \times \frac{365 - n + 1}{365} = \frac{365!}{(365 - n)! \, 365^n}$$

for $n = 1, 2, \ldots$.

(b) From (a) we find $\mathbb{P}(N > 23) \approx 0.492703$ and $\mathbb{P}(N > 22) \approx 0.524305$. Hence, the smallest n for which $\mathbb{P}(N > n) < 1/2$ is $n = 23$.

(c) $\mathbb{E}[N] = \sum_{i=1}^{\infty} n \, \mathbb{P}(N = n) = \sum_{i=1}^{\infty} n \{ \mathbb{P}(N > n - 1) - \mathbb{P}(N > n) \} \approx 24.6166$. Alternatively,

$$\mathbb{E}[N] = \sum_{n=0}^{\infty} \mathbb{P}(N > n) = \sum_{n=0}^{\infty} \frac{365!}{(365 - n)! \, 365^n} \approx 24.6166 \, .$$

1.6   Let $\mathbf{Z} = (U, V)^T$ and $\mathbf{X} = (X, Y)^T$. Then $\mathbf{Z} = A\mathbf{X}$, where A is the matrix

$$A = \begin{pmatrix} \sin \alpha & -\cos \alpha \\ \cos \alpha & \sin \alpha \end{pmatrix} .$$

Note that A is an *orthogonal* matrix (that is, $A^T A = I$) and, in particular, that its determinant, $|A|$, is equal to 1. The pdf of the random vector $\mathbf{X}$ is

$$f_{\mathbf{X}}(\mathbf{x}) = f_X(x) f_Y(y) = \frac{1}{2\pi} e^{-\frac{1}{2}(x^2 + y^2)} = \frac{1}{2\pi} e^{-\frac{1}{2}\mathbf{x}^T \mathbf{x}} .$$

By the transformation formula (1.19), the pdf of $\mathbf{Z}$ is found to be

$$f_{\mathbf{Z}}(\mathbf{z}) = \frac{f_{\mathbf{X}}(A^{-1}\mathbf{z})}{|A|} = f_{\mathbf{X}}(A^{-1}\mathbf{z}) = \frac{1}{2\pi} e^{-\frac{1}{2}\mathbf{z}^T (A^{-1})^T A^{-1} \mathbf{z}} .$$

Since A is orthogonal, A^{-1} is equal to its complement, A^T, and therefore the pdf of $\mathbf{Z}$ is

$$f_{\mathbf{Z}}(\mathbf{z}) = \frac{1}{2\pi} e^{-\frac{1}{2}\mathbf{z}^T A A^{-1} \mathbf{z}} = \frac{1}{2\pi} e^{-\frac{1}{2}\mathbf{z}^T \mathbf{z}} = \frac{1}{2\pi} e^{-\frac{1}{2}(u^2 + v^2)}$$

$$= \frac{1}{\sqrt{2\pi}} e^{-\frac{1}{2}u^2} \frac{1}{\sqrt{2\pi}} e^{-\frac{1}{2}v^2} .$$

In other words, U and V are independent and standard normally distributed.

1.7   First, note that $\mathbb{P}(X > x) = e^{-\lambda x}$ for all $x \geqslant 0$. Second, by the definition of conditional probability (1.3),

$$\mathbb{P}(X > t + s \,|\, X > t) = \frac{\mathbb{P}(\{X > t + s\} \cap \{X > t\})}{\mathbb{P}(X > t)} .$$

Since the event $\{X > t + s\}$ is a subset of $\{X > t\}$ for s and $t \geqslant 0$, their intersection is equal to $\{X > t + s\}$. Hence,

$$\mathbb{P}(X > t + s \mid X > t) = \frac{\mathbb{P}(X > t + s)}{\mathbb{P}(X > t)} = \frac{e^{-\lambda(s+t)}}{e^{-\lambda t}} = e^{-\lambda s},$$

which is equal to $\mathbb{P}(X > s)$, which had to be shown.

1.8   There are $2^3 = 8$ possible outcomes for the vector (X_1, X_2, X_3). Only three of these outcomes are such that the sum of the components is 2, namely $(1, 1, 0)$, $(1, 0, 1)$, and $(0, 1, 1)$. It follows that $\mathbb{P}(X_1 + X_2 + X_3 = 2) = \frac{1}{2}\frac{1}{3}\frac{3}{4} + \frac{1}{2}\frac{2}{3}\frac{1}{4} + \frac{1}{2}\frac{1}{3}\frac{1}{4} = \frac{1}{4}$ and that the conditional joint discrete pdf is given by

$$\mathbb{P}(X_1 = 1, X_2 = 1, X_3 = 0 \mid X_1 + X_2 + X_3 = 2)$$
$$= \frac{\mathbb{P}(X_1 = 1, X_2 = 1, X_3 = 0)}{\mathbb{P}(X_1 + X_2 + X_3 = 2)} = \frac{\frac{1}{2}\frac{1}{3}\left(1 - \frac{1}{4}\right)}{\frac{1}{4}} = \frac{1}{2},$$

$$\mathbb{P}(X_1 = 1, X_2 = 0, X_3 = 1 \mid X_1 + X_2 + X_3 = 2) = \frac{\frac{1}{2}\frac{2}{3}\frac{1}{4}}{\frac{1}{4}} = \frac{1}{3},$$

$$\mathbb{P}(X_1 = 0, X_2 = 1, X_3 = 1 \mid X_1 + X_2 + X_3 = 2) = \frac{\frac{1}{2}\frac{1}{3}\frac{1}{4}}{\frac{1}{4}} = \frac{1}{6}.$$

1.9   Although the expectations and variances can be derived by direct summation or integration, the derivations can often be simplified through the use of transform techniques.

    In particular, let X be a *nonnegative* and *integer-valued* random variable. The *probability generating function* of X is the function $G : [-1, 1] \to [0, 1]$ defined by

$$G(z) = \mathbb{E}[z^X] = \sum_{x=0}^{\infty} z^x \, \mathbb{P}(X = x) \,.$$

Differentiating $G(z)$ twice with respect to z gives

$$G'(z) = \frac{d\,\mathbb{E}[z^X]}{dz} = \mathbb{E}\left[\frac{d}{dz} z^X\right] = \mathbb{E}[X z^{X-1}] \,,$$

$$G''(z) = \frac{d\,\mathbb{E}[X z^{X-1}]}{dz} = \mathbb{E}[X(X-1) z^{X-2}] \,.$$

In particular, $\mathbb{E}[X] = G'(1)$ and $\text{Var}(X) = G''(1) + G'(1) - (G'(1))^2$.

    Similarly, the *moment generating function* of a random variable X is defined as the function $M : I \to [0, \infty)$ given by

$$M(s) = \mathbb{E}[e^{sX}] \,.$$

Here, I is an open interval containing 0 on which the above expectation is finite. If $\mathbb{E}[X^n]$ exists, then M is n times differentiable and

$$\mathbb{E}[X^n] = M^{(n)}(0) \qquad (n\text{-th derivative at } 0).$$

(a) If $X \sim \text{Bin}(n, p)$, then X can be written as $X = X_1 + \cdots + X_n$, where $\{X_i\} \sim_{iid} \text{Ber}(p)$ (see **Example 1.4**). Thus,

$$\mathbb{E}[X] = \mathbb{E}[X_1 + \cdots + X_n] = n \, \mathbb{E}[X_1] = n \, p \, .$$

Similarly,

$$\text{Var}(X) = n \, \text{Var}(X_1) = n \left\{ \mathbb{E}[X_1^2] - (\mathbb{E}[X_1])^2 \right\} = n \, (p - p^2) = n \, p \, (1 - p) \, .$$

(b) The probability generating function of $X \sim G(p)$ is given by

$$G(z) = \sum_{x=1}^{\infty} z^x p(1 - p)^{x-1} = z \, p \sum_{k=0}^{\infty} (z(1 - p))^k = \frac{z \, p}{1 - z \, (1 - p)} \, .$$

Consequently,

$$G'(z) = \frac{p}{(1 - z \, (1 - p))^2} \quad \text{and} \quad G''(z) = \frac{2(1 - p)p}{(1 - z \, (1 - p))^3} \, .$$

The expectation is therefore

$$\mathbb{E}[X] = G'(1) = \frac{1}{p}$$

and the variance is

$$\text{Var}(X) = G''(1) + G'(1) - (G''(1))^2 = \frac{2(1 - p)}{p^2} + \frac{1}{p} - \frac{1}{p^2} = \frac{1 - p}{p^2} \, .$$

(c) The probability generating function of $X \sim \text{Poi}(\lambda)$ is given by

$$G(z) = \sum_{x=0}^{\infty} z^x e^{-\lambda} \frac{\lambda^x}{x!} = e^{-\lambda} \sum_{x=0}^{\infty} \frac{(z\lambda)^x}{x!} = e^{-\lambda} e^{z\lambda} = e^{-\lambda(1-z)} \, .$$

Thus, $G'(z) = \lambda \, G(z)$ and $G''(z) = \lambda^2 \, G(z)$. It follows that $\mathbb{E}[X] = G'(1) = \lambda$ and $\text{Var}(X) = G''(1) + G'(1) - (G'(1))^2 = \lambda^2 + \lambda - \lambda^2 = \lambda$.

(d) If $X \sim \text{U}(\alpha, \beta)$, then $X = a + (b - a)Z$, with $Z \sim \text{U}(0, 1)$. Now, $\mathbb{E}[Z] = 1/2$ and $\text{Var}(Z) = \int_0^1 x^2 \, dx - (1/2)^2 = 1/12$. Thus, $\mathbb{E}[X] = a + (b - a)/2 = (a + b)/2$ and $\text{Var}(X) = (b - a)^2 \, \text{Var}(Z) = (b - a)^2/12$.

(e) The moment generating function of $X \sim \text{Gamma}(\alpha, \lambda)$ is given by

$$
\begin{aligned}
M(s) = \mathbb{E}[e^{sX}] &= \int_0^\infty \frac{e^{-\lambda x} \lambda^\alpha x^{\alpha-1}}{\Gamma(\alpha)} e^{sx} \, dx \\
&= \left(\frac{\lambda}{\lambda - s} \right)^\alpha \int_0^\infty \frac{e^{-(\lambda-s)x} (\lambda - s)^\alpha x^{\alpha-1}}{\Gamma(\alpha)} \, dx \\
&= \left(\frac{\lambda}{\lambda - s} \right)^\alpha \, .
\end{aligned}
\tag{11.1}
$$

As a consequence,

$$\mathbb{E}[X] = M'(0) = \frac{\alpha}{\lambda} \left(\frac{\lambda}{\lambda - s} \right)^{\alpha+1} \Bigg|_{s=0} = \frac{\alpha}{\lambda}$$

and

$$\mathbb{E}[X^2] = M''(0) = \frac{\alpha(\alpha+1)}{\lambda^2} \left(\frac{\lambda}{\lambda-s}\right)^{\alpha+1}\bigg|_{s=0} = \frac{\alpha(\alpha+1)}{\lambda^2},$$

so that

$$\mathrm{Var}(X) = \frac{\alpha(\alpha+1)}{\lambda^2} - \left(\frac{\alpha}{\lambda}\right)^2 = \frac{\alpha}{\lambda^2}.$$

(f) The results for $X \sim \mathrm{Exp}(\lambda)$ follow from the $\mathrm{Gamma}(1, \lambda)$ case.

(g) The moment generation function of $Z \sim \mathrm{N}(0,1)$ is

$$M(s) = \mathbb{E}[e^{sZ}] = \int_{-\infty}^{\infty} e^{sz} \frac{1}{\sqrt{2\pi}} e^{-z^2/2} dz = e^{s^2/2} \int_{-\infty}^{\infty} \frac{1}{\sqrt{2\pi}} e^{-(z-s)^2/2} dz = e^{s^2/2}.$$

It follows that $\mathbb{E}[Z^2] = M''(0) = 1$ and $\mathbb{E}[Z] = 0$ (or by symmetry), so that $\mathrm{Var}(Z) = 1$. For $X \sim \mathrm{N}(\mu, \sigma^2)$, write $X = \mu + \sigma Z$, whence $\mathbb{E}[X] = \mu + \sigma \mathbb{E}[Z] = \mu$ and $\mathrm{Var}(X) = \sigma^2 \mathrm{Var}(Z) = \sigma^2$.

(h) Let $X \sim \mathrm{Beta}(\alpha, \beta)$. Then,

$$\begin{aligned}
\mathbb{E}[X] &= \int_0^1 \frac{\Gamma(\alpha+\beta)}{\Gamma(\alpha)\Gamma(\beta)} x^\alpha (1-x)^{\beta-1} dx \\
&= \frac{\Gamma(\alpha+\beta)\Gamma(\alpha+1)}{\Gamma(\alpha+\beta+1)\Gamma(\alpha)} \int_0^1 \frac{\Gamma(\alpha+\beta+1)}{\Gamma(\alpha+1)\Gamma(\beta)} x^\alpha (1-x)^{\beta-1} dx \\
&= \frac{\Gamma(\alpha+\beta)\Gamma(\alpha+1)}{\Gamma(\alpha+\beta+1)\Gamma(\alpha)} \\
&= \frac{\alpha}{\alpha+\beta},
\end{aligned}$$

using the fact that $\Gamma(u+1) = u\Gamma(u)$. Similarly, for the second moment we have

$$\mathbb{E}[X^2] = \frac{\Gamma(\alpha+\beta)\Gamma(\alpha+2)}{\Gamma(\alpha+\beta+2)\Gamma(\alpha)} = \frac{(\alpha+1)\alpha}{(\alpha+\beta+1)(\alpha+\beta)},$$

so that

$$\mathrm{Var}(X) = \frac{(\alpha+1)\alpha}{(\alpha+\beta+1)(\alpha+\beta)} - \left(\frac{\alpha}{\alpha+\beta}\right)^2 = \frac{\alpha\beta}{(\alpha+\beta)^2(1+\alpha+\beta)}.$$

(i) Let $X \sim \mathrm{Weib}(\alpha, \lambda)$. Then,

$$\begin{aligned}
\mathbb{E}[X] &= \int_0^\infty \alpha(\lambda x)^\alpha e^{-(\lambda x)^\alpha} dx \\
&= \frac{1}{\lambda} \int_0^\infty e^{-y} y^{1/\alpha} dy \quad \text{(change of variable: } y = (\lambda x)^\alpha\text{)} \\
&= \frac{\Gamma(1+1/\alpha)}{\lambda} = \frac{\Gamma(1/\alpha)}{\alpha\lambda}.
\end{aligned}$$

Similarly,

$$
\begin{aligned}
\mathbb{E}[X^2] &= \frac{\alpha}{\lambda} \int_0^\infty (\lambda x)^{\alpha+1} e^{-(\lambda x)^\alpha}\, dx \\
&= \int_0^\infty e^{-y} y^{2/\alpha}\, dy \\
&= \frac{\Gamma(1+2/\alpha)}{\lambda} = \frac{2\Gamma(2/\alpha)}{\alpha},
\end{aligned}
$$

so that $\mathrm{Var}(X) = \frac{2\,\Gamma(2/\alpha)}{\alpha} - \left(\frac{\Gamma(1/\alpha)}{\alpha\lambda}\right)^2$.

1.10   Let D be the set $\{(x,y) : 0 \leqslant y \leqslant 1,\ 0 \leqslant x \leqslant 1\}$. The density has to integrate to 1 on D, that is,

$$
\iint_D c\,x\,y\,dx\,dy = c \int_0^1 \int_0^x x\,y\,dy\,dx = c\,\frac{1}{8} = 1,
$$

which gives $c = 8$. Next, let $A = \{(x,y) \in D : x + 2\,y \leqslant 1\}$. Then, $\mathbb{P}(X + 2Y \leqslant 1) = \mathbb{P}((X,Y) \in A) = \iint_A f(x,y)\,dx\,dy = \int_0^{1/3} \int_y^{1-2y} 8\,x\,y\,dx\,dy = \frac{5}{81}$.

1.11 (a) For any $z \geqslant 0$, $\mathbb{P}(\min(X,Y) > z) = \mathbb{P}(X > z, Y > z) = \mathbb{P}(X > z)\,\mathbb{P}(Y > z) = e^{-\lambda z}\,e^{-\mu z} = e^{-(\lambda+\mu)z}$. In other words, $\min(X,Y) \sim \mathrm{Exp}(\lambda + \mu)$.

(b) For any $z \geqslant 0$ and $h > 0$,

$$
\begin{aligned}
\mathbb{P}(X < Y \mid \min(X,Y) \in [z, z+h]) &= \frac{\int_z^{z+h} \int_x^\infty \lambda\mu\, e^{-\lambda x} e^{-\mu y}\, dy\, dx}{\int_z^{z+h} (\lambda+\mu) e^{-(\lambda+\mu)u}\, du} \\
&= \frac{\lambda}{\lambda+\mu}.
\end{aligned}
$$

Hence, $\mathbb{P}(X < Y \mid \min(X,Y)) = \lambda/(\lambda+\mu)$.

1.12   For simplicity of notation we write $\mathbb{E}[Z] = \mu_Z$ and $\mathrm{Var}(Z) = \sigma_Z^2$ for a generic random variable Z.

1. $\mathrm{Var}(X) = \mathbb{E}[X - \mu_X]^2 = \mathbb{E}[X^2] - 2\,\mathbb{E}[X\,\mu_X] + \mu_X^2 = \mathbb{E}[X^2] - \mu_X^2$.

2. Let $Y = a\,X + b$. By the linearity of the expectation, $\mu_Y = a\,\mu_X + b$. Hence, $\mathrm{Var}(Y) = \mathbb{E}[\{a + bX - (a\,\mu_X + b)\}^2] = \mathbb{E}[a^2\,(X - \mu_X)^2] = a^2\,\mathrm{Var}(X)$.

3. $\mathrm{Cov}(X,Y) = \mathbb{E}[(X - \mu_X)(Y - \mu_Y)] = \mathbb{E}[XY - X\,\mu_Y - Y\,\mu_X + \mu_X\,\mu_Y] = \mathbb{E}[XY] - \mu_X\,\mu_Y$.

4. $\mathrm{Cov}(X,Y) = \mathbb{E}[(X - \mu_X)(Y - \mu_Y)] = \mathbb{E}[(Y - \mu_Y)(X - \mu_X)] = \mathrm{Cov}(Y,X)$.

5. $\mathrm{Cov}(aX + bY, Z) = \mathbb{E}[(aX + bY)Z] - \mathbb{E}[aX + bY]\,\mathbb{E}[Z] = a\,\mathbb{E}[XZ] - a\,\mathbb{E}[XZ] + b\,\mathbb{E}[YZ] - b\,\mathbb{E}[YZ] = a\,\mathrm{Cov}(X,Z) + b\,\mathrm{Cov}(Y,Z)$.

6. $\mathrm{Cov}(X,X) = \mathbb{E}[(X - \mu_X)(X - \mu_X)] = \mathbb{E}[(X - \mu_X)^2] = \mathrm{Var}(X)$.

7. By Property 6 of **Table 1.4** (see Table 1.2): $\mathrm{Var}(X + Y) = \mathrm{Cov}(X + Y, X + Y)$. By Property 5: $\mathrm{Cov}(X+Y, X+Y) = \mathrm{Cov}(X,X) + \mathrm{Cov}(Y,Y) + \mathrm{Cov}(X,Y) + \mathrm{Cov}(Y,X) = \mathrm{Var}(X) + \mathrm{Var}(Y) + 2\,\mathrm{Cov}(X,Y)$, where in the last equation Properties 4 and 6 are used.

8. If X and Y are independent, then $\mathbb{E}[XY] = \mu_X\,\mu_Y$. Therefore, $\mathrm{Cov}(X,Y) = 0$ follows immediately from Property 3 of Table 1.2.

1.13 Let a be an arbitrary real number and denote the standard deviations of X and Y by σ_X and σ_Y. The variance of $-aX + Y$ is always nonnegative. Thus, using the properties of covariance and variance in Table 1.2,

$$\mathrm{Var}(-aX + Y) = a^2\sigma_X^2 + \sigma_Y^2 - 2a\,\mathrm{Cov}(X,Y) \geqslant 0 \,.$$

Now take $a = \mathrm{Cov}(X,Y)/\sigma_X^2$. Then, after rearranging, we obtain

$$\{\mathrm{Cov}(X,Y)\}^2 \leqslant \sigma_X^2\sigma_Y^2 \,,$$

which is equivalent to

$$-1 \leqslant \varrho(X,Y) \leqslant 1 \,.$$

1.14 This requires a knowledge of measure theory. Let F be a cdf. Then there exists a unique probability measure μ on the Borel σ-algebra such that $\mu((-\infty, x]) = F(x)$ for all x. If we let $\Omega = \mathbb{R}$, $\mathbb{P} = \mu$, and define $X : \Omega \to \mathbb{R}$ as the identity function (that is, $X(\omega) = \omega$), then, by construction, $\mathbb{P}(X \leqslant x) = \mathbb{P}(\{\omega : X(\omega) \leqslant x\}) = \mathbb{P}((-\infty, x]) = \mu((-\infty, x]) = F(x)$. Hence, X is a random variable with cdf F.

1.15 By virtue of Property 6 of Table 1.2 and repetitive application of Property 5, we have

$$\mathrm{Var}\left(\sum_{i=1}^{n} X_i\right) = \mathrm{Cov}(X_1 + X_2 + \cdots + X_n, \; X_1 + X_2 + \cdots + X_n)$$

$$= \sum_{i=1}^{n} \mathrm{Cov}(X_i, X_i) + 2\sum_{i<j} \mathrm{Cov}(X_i, X_j)$$

$$= \sum_{i=1}^{n} \mathrm{Var}(X_i) + 2\sum_{i<j} \mathrm{Cov}(X_i, X_j) \,.$$

1.16 Since Σ is a covariance matrix, the Cholesky Square Root Method (see **Section A.1**) can be used to factorize Σ into $\Sigma = BB^T$, where B is a lower-triangular matrix. Then, $\mathbf{u}^T\Sigma\mathbf{u} = (B^T\mathbf{u})^T(B^T\mathbf{u}) = \|B^T\mathbf{u}\|^2 \geqslant 0$ for any $\mathbf{u}$. Hence, Σ is a positive semidefinite matrix.

1.17 Take the derivative of the cdf F of Y to get the pdf f as follows:

$$f(x) = \frac{d}{dx}F(x) = \frac{d}{dx}\left(1 - \sum_{k=0}^{n-1}\frac{e^{-\lambda x}(\lambda x)^k}{k!}\right) = -\sum_{k=0}^{n-1}\frac{d}{dx}\frac{e^{-\lambda x}(\lambda x)^k}{k!}$$

$$= \sum_{k=0}^{n-1}\left\{\frac{e^{-\lambda x}\lambda(\lambda x)^k}{k!} - \frac{e^{-\lambda x}k\lambda(\lambda x)^{k-1}}{k!}\right\}$$

$$= \sum_{k=0}^{n-1}\frac{e^{-\lambda x}\lambda(\lambda x)^k}{k!} - \sum_{k=0}^{n-2}\frac{e^{-\lambda x}\lambda(\lambda x)^k}{k!}$$

$$= \frac{e^{-\lambda x}\lambda(\lambda x)^{(n-1)}}{(n-1)!} \,,$$

which is the pdf of the Gamma(n, λ) distribution.

1.18 Since $M = \min\{X_1, \ldots, X_n\}$, we can write

$$\mathbb{P}(M > m) = \mathbb{P}(X_1 > m, \ldots, X_n > m) = \mathbb{P}(X_1 > m)\cdots\mathbb{P}(X_n > m) = (1-m)^n \,,$$

due to the independence assumption. Therefore, for each $n \geq 1$, the cdf of M is given by

$$F(m) = \mathbb{P}(M \leq m) = 1 - (1-m)^n, \quad 0 \leq m \leq 1,$$

and the pdf is

$$f(m) = \frac{dF(m)}{dm} = n(1-m)^{n-1}, \quad 0 \leq m \leq 1.$$

1.19 (a) Applying the transformation rule **(1.16)** or **(1.20)** yields

$$f_Y(y) = f_X(\ln y)\frac{d}{dy}\ln y = \frac{e^{-\frac{1}{2}(\ln y)^2}}{y\sqrt{2\pi}}, \quad y > 0.$$

(b) The expectation of Y follows from

$$\mathbb{E}[Y] = \mathbb{E}[e^X] = \int_{-\infty}^{\infty} e^x \frac{e^{-\frac{1}{2}x^2}}{\sqrt{2\pi}} = e^{\frac{1}{2}} \int_{-\infty}^{\infty} \frac{e^{-\frac{1}{2}(x-1)^2}}{\sqrt{2\pi}} = e^{\frac{1}{2}}.$$

1.20 (a) We are given that $f_X(x) = 1$ on $x \in (0,1)$ and

$$f_{Y\,|\,X}(y\,|\,x) = \begin{cases} \frac{1}{x} & 0 < y < x < 1 \\ 0 & \text{otherwise.} \end{cases}$$

Therefore, the joint pdf is given by

$$f_{XY}(x,y) = f_X(x)\,f_{Y\,|\,X}(y\,|\,x) = \begin{cases} \frac{1}{x} & 0 < y < x < 1 \\ 0 & \text{otherwise.} \end{cases}$$

(b) $f_Y(y) = \int_{-\infty}^{\infty} f_{XY}(x,y)\,dx = \int_y^1 \frac{1}{x}\,dx = -\ln(y)$ for $0 < y < 1$.

(c) $f_{X\,|\,Y}(x\,|\,y) = \frac{f_{XY}(x,y)}{f_Y(y)} = \frac{-1}{x\ln(y)}$ for $y < x < 1$ (and 0 otherwise).

(d) $\mathbb{E}[X\,|\,Y = y] = \int_y^1 x\frac{-1}{x\ln(y)}\,dx = \frac{y-1}{\ln(y)}$.

(e) $\mathbb{E}[X] = \int_0^1 x\,dx = 1/2$ and $\mathbb{E}[Y] = \mathbb{E}[\mathbb{E}[Y\,|\,X]] = \int_0^1 \int_0^x \frac{y}{x}\,dy\,dx = \int_0^1 \frac{x}{2}\,dx = \frac{1}{4}$.

1.21 (a)

$$\begin{aligned} \mathbb{P}(N_2 = 1, N_3 = 4, N_5 = 5) &= \mathbb{P}(N[0,2] = 1, N(2,3] = 3, N(3,5] = 1) \\ &= \mathbb{P}(N[0,2] = 1)\,\mathbb{P}(N(2,3] = 3)\,\mathbb{P}(N(3,5] = 1) \\ &= \frac{e^{-4}4^1}{1!}\frac{e^{-2}2^3}{3!}\frac{e^{-4}4^1}{1!} \\ &= \frac{e^{-10}64}{3} \approx 9.68 \cdot 10^{-4}. \end{aligned}$$

(b) $\mathbb{P}(N_4 = 4\,|\,N_2 = 1, N_3 = 2) = \mathbb{P}(N(3,4] = 1) = 2\,e^{-2} \approx 0.27$.

(c) $\mathbb{E}[N_4\,|\,N_2 = 2] = 2 + \mathbb{E}[N(2,4]] = 2 + 4 = 6$.

(d)

$$\begin{aligned} \mathbb{P}(N[2,7] = 4, N[3,8] = 6) &= \sum_{k=0}^{4} \mathbb{P}(N[2,3] = k, N(3,7] = 4-k, N(7,8] = 2+k) \\ &= \sum_{k=1}^{4} \frac{e^{-2}2^k}{k!}\frac{e^{-8}8^{4-k}}{(4-k)!}\frac{e^{-2}2^{2+k}}{(2+k)!} \\ &= e^{-12}\frac{82952}{135} \approx 3.77 \cdot 10^{-3}. \end{aligned}$$

(e) $\mathbb{E}[N[4,6] \mid N[1,5] = 3] = \mathbb{E}[N[4,5] \mid N[1,5] = 3] + \mathbb{E}[N(5,6]] = \frac{3}{4} + 2 = 2.75$.

1.22   First, note that $\binom{n}{k} = \frac{n!}{(n-k)!\,k!}$ is a polynomial in n of degree k with leading coefficient $(k!)^{-1}$. Therefore, by applying L'Hôpital's rule k times, we find

$$\lim_{n\to\infty} \binom{n}{k} / n^k = (k!)^{-1} .$$

Hence,

$$\lim_{n\to\infty} \binom{n}{k} \frac{(\lambda t)^k}{n^k} \left(1 - \frac{\lambda t}{n}\right)^{n-k} = (\lambda t)^k \times \lim_{n\to\infty} \frac{\binom{n}{k}}{n^k} \times \lim_{n\to\infty} \left(1 - \frac{\lambda t}{n}\right)^{n-k}$$
$$= \frac{(\lambda t)^k}{k!} e^{-\lambda t} ,$$

where we have used the definition of the exponential function, the fact that k is fixed, and that $n - k \to \infty$ for $n \to \infty$.

1.23 (a) The $X_1, X_2, \ldots$ form a Bernoulli process with success parameter $p = \lambda h$. The time of first success, U_1, is therefore $G(p)$ distributed. Starting anew from time U_1, the number of trials required to the next success, $U_2 - U_1$, is $G(p)$ distributed as well and is independent of U_1, and so on.

(b) By the geometric formula $\mathbb{P}(U_1 > n) = (1-p)^n$, which is approximated by $(1 - \lambda t/n)^n$ and converges to $e^{-\lambda t}$ as $n \to \infty$.

1.24   From the properties of conditional probability and $N_t \sim \text{Poi}(\lambda)$, we find that

$$\mathbb{P}(N_u = j \mid N_t = n) = \frac{\mathbb{P}(N_u = j, \, N_t = n)}{\mathbb{P}(N_t = n)} = \frac{\mathbb{P}(N_u = j, \, N_{t-u} = n - j)}{\mathbb{P}(N_t = n)}$$
$$= \frac{\mathbb{P}(N_u = j) \, \mathbb{P}(N_{t-u} = n - j)}{\mathbb{P}(N_t = n)} ,$$

by using the property of the Poisson arrival counting process that the number of arrivals in nonoverlapping regions are independent. Hence,

$$\mathbb{P}(N_u = j \mid N_t = n) = \frac{e^{-\lambda u} \dfrac{(\lambda u)^j}{j!} \; e^{-\lambda(t-u)} \dfrac{(\lambda(t-u))^{n-j}}{(n-j)!}}{e^{-\lambda t} \dfrac{(\lambda t)^n}{n!}}$$
$$= \binom{n}{j} \left(\frac{u}{t}\right)^j \left(1 - \frac{u}{t}\right)^{n-j} .$$

1.25   Write $X_n = 2 \sum_{i=1}^n U_i - n$, were the $\{U_i\}$ are iid and $\text{Ber}(p)$ distributed. Thus, $X_n = 2Y - n$, with $Y \sim \text{Bin}(n, p)$. Then,

(a) $\mathbb{P}(X_n = 2i - n) = \mathbb{P}(Y = i) = \binom{n}{i} p^i (1-p)^{n-i}$, $i = 0, 1, \ldots, n$.

(b) $\mathbb{E}[X_n] = \mathbb{E}[2Y - n] = 2np - n = 2np - np - nq = n(p - q)$, with $q = 1 - p$.

(c) $\text{Var}(X_n) = 4\,\text{Var}(Y) = 4\,npq$.

1.26 (a) $\mathbb{P}(X_1 = 2) = (\pi P)(3) = (0.29, 0.43, 0.28)(3) = 0.28$.

(b) $\mathbb{P}(X_2 = 2) = (\pi P^2)(3) = (0.2870, 0.3970, 0.3160)(3) = 0.316$.

(c) $\mathbb{P}(X_3 = 2 \mid X_0 = 0) = (P^3)(1,3) = 0.276$.

(d)

$$\mathbb{P}(X_0 = 1 \mid X_1 = 2) = \frac{\mathbb{P}(X_0 = 1, X_1 = 2)}{\mathbb{P}(X_1 = 2)}$$

$$= \frac{\mathbb{P}(X_1 = 2 \mid X_0 = 1)\,\mathbb{P}(X_0 = 1)}{\mathbb{P}(X_1 = 2)}$$

$$= \frac{0.2 \times 0.5}{0.28} = 0.3571 .$$

(e) $\mathbb{P}(X_1 = 1, X_3 = 1) = \mathbb{P}(X_3 = 1 \mid X_1 = 1)\,\mathbb{P}(X_1 = 1) = \mathbb{P}(X_2 = 1 \mid X_0 = 1)\,\mathbb{P}(X_1 = 1) = P^2(2,2) \times (\pi P)(2) = 0.34 \times 0.43 = 0.1462$.

1.27   Let X_n be the number of fleas on Spot at time n. Then $\{X_n, n = 0, 1, \ldots\}$ is a Markov chain starting in b, with the following transition graph.

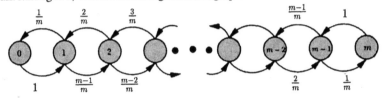

The stationary distribution is found from the detailed balance equations (**1.43**), which in this case are as follows:

$$\pi_0 = \frac{1}{m}\pi_1$$

$$\frac{m-1}{m}\pi_1 = \frac{2}{m}\pi_2$$

$$\frac{m-2}{m}\pi_2 = \frac{3}{m}\pi_3$$

$$\vdots$$

$$\frac{1}{m}\pi_m = \frac{m-1}{m}\pi_{m-1} .$$

It follows that for $k = 0, 1, \ldots, m$,

$$\pi_k = \frac{m!/(m-k)!}{k!}\pi_0 = \binom{m}{k}\pi_0 ,$$

From normalization,

$$\sum_{k=0}^{m}\binom{m}{k}\pi_0 = (1+1)^m\pi_0 = 2^m\pi_0 = 1 ,$$

we find that $\pi_0 = 2^{-m}$. In other words, the steady-state number of fleas on Spot has a $\text{Bin}(m, 1/2)$ distribution.

1.28   By examination of the one-step transition matrix, we conclude that states 1, 3, and 5 *communicate*. Thus, the set $A = \{1, 3, 5\}$ is an equivalence class. The set $B = \{2, 4\}$ is another equivalence class. Notice that once B is entered, the Markov chain cannot go back to set A. In other words, B is a closed set and hence the states in B are *recurrent*. States 2 and 4 are positive recurrent and aperiodic, because the probability of remaining in either state 2 or 4 and not jumping to the other is positive.

1.29   Let N be the time the game is finished and X_n be the position at time n. Then $\{X_n\}$ is a Markov chain with the following transition matrix

	1	2	3	4	5	6	7	8	9	10	11	12	13	14	15	16	17	18	19	20
1	0	1/6	0	1/6	1/6	1/6	1/6	0	0	0	1/6	0	0	0	0	0	0	0	0	0
2	0	0	0	1/6	1/6	1/6	1/6	1/6	0	0	1/6	0	0	0	0	0	0	0	0	0
3	0	0	0	0	0	0	0	0	0	0	1	0	0	0	0	0	0	0	0	0
4	0	0	0	0	1/6	1/6	1/6	1/6	1/6	1/6	0	0	0	0	0	0	0	0	0	0
5	0	0	0	0	0	1/6	1/6	1/6	1/6	1/6	1/6	1/6	0	0	0	0	0	0	0	0
6	0	0	0	0	0	0	1/6	1/6	1/6	1/6	1/6	1/6	0	0	0	0	0	0	0	0
7	0	0	0	1/6	0	0	0	1/6	1/6	1/6	1/6	1/6	0	0	0	0	0	0	0	0
8	0	0	0	1/6	0	0	0	0	1/6	1/6	1/6	1/6	0	1/6	0	0	0	0	0	0
9	0	0	0	1/6	0	0	0	0	0	1/6	1/6	1/6	0	1/6	0	0	0	0	1/6	0
10	0	0	0	1/6	0	0	0	0	0	0	1/6	1/6	0	1/6	0	1/6	0	0	1/6	0
11	0	0	0	1/6	0	0	0	0	0	1/6	0	1/6	0	1/6	0	1/6	0	1/6	0	0
12	0	0	0	1/6	0	0	0	0	0	1/6	0	0	0	1/6	0	1/6	0	1/6	1/6	0
13	0	0	0	1	0	0	0	0	0	0	0	0	0	0	0	0	0	0	0	0
14	0	0	0	0	0	0	0	0	0	1/6	0	0	0	0	0	1/6	0	1/6	1/3	1/6
15	0	0	0	0	0	0	0	0	0	0	0	0	0	0	0	0	0	0	1	0
16	0	0	0	0	0	0	0	0	0	1/6	0	0	0	0	0	0	0	1/6	1/6	1/2
17	0	0	0	0	0	0	0	0	0	1	0	0	0	0	0	0	0	0	0	0
18	0	0	0	0	0	0	0	0	0	0	0	0	0	0	0	0	0	0	1/6	5/6
19	0	0	0	0	0	0	0	0	0	0	0	0	0	0	0	0	0	0	0	1
20	0	0	0	0	0	0	0	0	0	0	0	0	0	0	0	0	0	0	0	1

Rows 3, 13, 15, and 17 are irrelevant since the Markov chain will never visit these states when starting from 1. Note that state 20 is an absorbing state. Thus, we have

$$\mathbb{P}(N \leqslant n) = \mathbb{P}(X_n = 20) = P^n(1, 20)$$

and

$$\mathbb{E}[N] = \sum_{n=0}^{\infty} (1 - P^n(1, 20)) \approx 6.90 .$$

1.30   Let Y_n be the number of umbrellas at the place of the n-th arrival. Then $\{Y_n, n \in \mathbb{N}\}$ is a Markov chain with state space $E = \{0, 1, 2, 3\}$ and the following one-step transition matrix.

	0	1	2	3
0	0	0	0	1
1	0	0	$1 - p$	p
2	0	$1 - p$	p	0
3	$1 - p$	p	0	0

The Markov chain is irreducible and aperiodic and the limiting distribution is the solution to $\pi P = \pi$ for $\pi_i \geqslant 0$, with $\sum_i \pi_i = 1$. Since the Markov chain is reversible, the detailed balance equations hold. The solution of the detailed balance equations is $\pi = (1 - p, 1, 1, 1)/(4 - p)$. Thus, $\lim_{n \to \infty} \mathbb{P}(Y_n = 0) = \frac{1-p}{4-p}$ is the probability that Ms. Brum goes to a place without any umbrella. Therefore, the limiting probability that it rains and no umbrella is available is $p \times \frac{1-p}{4-p}$.

1.31   The Markov chain is reversible. Solve the detailed balance equations to find $\pi = \frac{1}{17}(2, 2, 2, 3, 4, 4)$.

1.32   The Q-matrix of $X = \{X_t, t \geqslant 0\}$ with state space $\{0, 1, 2, \ldots\}$ is given by

$$Q = \begin{pmatrix} -a & a & 0 & 0 & 0 & \cdots \\ b & -b-a & a & 0 & 0 & \cdots \\ 0 & 2b & -2b-a & a & 0 & \cdots \\ \vdots & \vdots & & \ddots & \ddots & \cdots \end{pmatrix} .$$

Since X is a recurrent and irreducible Markov jump process, the solution of $\pi Q = 0$ with $\sum_i \pi_i = 1$ gives the limiting distribution π. Alternatively, since the process is reversible, we can use the detailed balance equations $\pi_i q_{ij} = \pi_j q_{ij}, i, j \in \{0, 1, 2, \ldots\}$, to find that the limiting distribution satisfies

$$\pi_1 = \frac{q_{01}}{q_{10}} \pi_0 = \frac{a}{b} \pi_0$$

$$\pi_2 = \frac{a}{2\,b} \pi_1 = \frac{a^2}{2!\,b^2} \pi_0$$

$$\pi_3 = \frac{a}{3\,b} \pi_2 = \frac{a^3}{3!\,b^3} \pi_0$$

$$\vdots$$

In general, we have the recursion $\pi_n = \frac{a}{n\,b} \pi_{n-1}$, $n = 1, 2, \ldots$, from which it follows that

$$\pi_n = \frac{(a/b)^n}{n!} \pi_0, \quad n = 0, 1, \ldots.$$

Imposing the normalizing condition $\sum_{i=0}^{\infty} \pi_i = 1$ yields

$$\pi_0 \sum_{i=0}^{\infty} \frac{(a/b)^n}{n!} = \pi_0 \, e^{a/b} = 1 \,.$$

Hence, the limiting distribution is given by

$$\lim_{t \to \infty} \mathbb{P}(X_t = n \mid X_0 = i) = \pi_n = \frac{(a/b)^n}{n!} e^{-(a/b)}, \quad n = 0, 1, 2, \ldots,$$

which is the Poisson density with parameter a/b.

1.33   Taking the derivative of $\mathbf{a}^T \mathbf{x}$ with respect to x_i gives, for each $i = 1, \ldots, n$,

$$\frac{\partial \, \mathbf{a}^T \mathbf{x}}{\partial x_i} = \frac{\partial (a_1 x_1 + \cdots a_n x_n)}{\partial x_i} = a_i \,.$$

In vector notation this is written as $\nabla_{\mathbf{x}}(\mathbf{a}^T \mathbf{x}) = \mathbf{a}$.

1.34   First, write $\mathbf{x}^T A \mathbf{x} = \sum_{k=1}^{n} \sum_{j=1}^{n} a_{jk} x_j x_k$. Taking the partial derivative with respect to x_i gives, for each $i = 1, \ldots, n$,

$$\frac{\partial \, \mathbf{x}^T A \mathbf{x}}{\partial x_i} = \sum_{k=1}^{n} a_{ik} x_k + \sum_{j=1}^{n} a_{ji} x_j = 2 \sum_{k=1}^{n} a_{ik} x_k \,,$$

where the last equality follows from the assumed symmetry of A. Summarized in column-vector notation, we obtain $\nabla_{\mathbf{x}}(\mathbf{x}^T A \mathbf{x}) = 2A\mathbf{x}$. For a nonsymmetrical A, the above procedure is readily modified to show that

$$\nabla_{\mathbf{x}}(\mathbf{x}^T A \mathbf{x}) = A\mathbf{x} + A^T \mathbf{x} \,.$$

1.35   Setting the derivative of the Lagrangian (with respect to p_i) equal to zero yields

$$1 + \ln(p_i) + \beta = 0, \quad i = 1, \ldots, n \,.$$

Hence, $p_i = e^{-(1+\beta)}$ is the same for all i. Since the $\{p_i\}$ sum up to 1, each p_i must be equal to $1/n$.

1.36   The Lagrangian function of

$$\min_{\boldsymbol{\alpha}} \ \frac{1}{2} \boldsymbol{\alpha}^T C \boldsymbol{\alpha} - \boldsymbol{\alpha}^T \mathbf{b}$$

$$\text{subject to:} \ \ \boldsymbol{\alpha} \geqslant \mathbf{0}$$

is

$$\mathcal{L}(\boldsymbol{\alpha}, \boldsymbol{\theta}) = \frac{1}{2} \boldsymbol{\alpha}^T C \boldsymbol{\alpha} - \boldsymbol{\alpha}^T \mathbf{b} - \boldsymbol{\theta}^T \boldsymbol{\alpha} \,,$$

with $\boldsymbol{\theta} \geqslant \mathbf{0}$. The corresponding Lagrange dual function is

$$\mathcal{L}^*(\boldsymbol{\theta}) = \inf_{\boldsymbol{\alpha} \geqslant \mathbf{0}} \ \mathcal{L}(\boldsymbol{\alpha}, \boldsymbol{\theta}) \,.$$

The minimizer of $\mathcal{L}(\boldsymbol{\alpha}, \boldsymbol{\theta})$ with respect to $\boldsymbol{\alpha}$ is found by taking the gradient of $\mathcal{L}(\boldsymbol{\alpha}, \boldsymbol{\theta})$ with respect to $\boldsymbol{\alpha}$, giving

$$C\boldsymbol{\alpha} - \mathbf{b} - \boldsymbol{\theta} = \mathbf{0} \,.$$

Substituting this back into $\mathcal{L}(\boldsymbol{\alpha}, \boldsymbol{\theta})$ yields

$$\mathcal{L}^*(\boldsymbol{\theta}) = -\frac{1}{2} \boldsymbol{\alpha}^T C \boldsymbol{\alpha} \,,$$

where $\boldsymbol{\alpha} = C^{-1}(\boldsymbol{\theta} + \mathbf{b})$. The dual problem is to maximize $\mathcal{L}^*(\boldsymbol{\theta})$ over $\boldsymbol{\theta} \geqslant \mathbf{0}$, which is equivalent to minimizing $\frac{1}{2} \boldsymbol{\alpha}^T C \boldsymbol{\alpha}$ over $\boldsymbol{\alpha}$ satisfying $C\boldsymbol{\alpha} \geqslant \mathbf{b}$. Finally, by writing $C = BB^T$ and replacing $B^T \boldsymbol{\alpha}$ with $\boldsymbol{\mu}$, the dual program becomes

$$\min_{\boldsymbol{\mu}} \ \frac{1}{2} \boldsymbol{\mu}^T \boldsymbol{\mu}$$

$$\text{subject to:} \ \ B\boldsymbol{\mu} \geqslant \mathbf{b} \,.$$

1.37 (a) The Lagrangian follows from direct combination of the objective function with the constraints, including the nonnegativity constraints.

(b) Solving (for fixed $\boldsymbol{\lambda}$, $\boldsymbol{\mu}$ and β)

$$\nabla_{\mathbf{p}} \mathcal{L}(\mathbf{p}, \boldsymbol{\lambda}, \boldsymbol{\mu}, \beta) = \mathbf{0} \,,$$

gives the set of equations

$$\mathbf{1} + \boldsymbol{\xi}(\mathbf{p}) - A^T \boldsymbol{\lambda} - \boldsymbol{\mu} + \beta \mathbf{1} = \mathbf{0} \,.$$

Solving for p_i gives

$$p_i = q_i \exp\left(-\beta - 1 + \mu_i + \sum_{j=1}^{m} \lambda_j \, a_{ji} \right), \quad i = 1, \dots, n \,. \tag{11.2}$$

(c) The KKT conditions on $\boldsymbol{\mu}$ are that $\mu_i \, p_i = 0$ for all $i = 1, \dots, n$. Since $p_i^* > 0$, it follows that $\mu_i = 0$, $i = 1, \dots, n$.

(d) The rest of the solution procedure is exactly the same as in **Example 1.20**. In particular, by summing up the $\{p_i\}$ we find that

$$e^\beta = \sum_{i=1}^{n} q_i \exp(-1 + \mathbf{1}^T A^T \boldsymbol{\lambda}) \,. \tag{11.3}$$

Substituting $p(\lambda, \beta)$ back into the Lagrangian gives

$$\mathcal{L}^*(\lambda, \beta) = -1 + \lambda^T b - \beta .\tag{11.4}$$

Next, solve the dual program $\max_{\lambda, \beta} \mathcal{L}^*(\lambda, \beta)$. Since β and λ are related via (11.3), this can be done by substituting the corresponding $\beta(\lambda)$ into (11.4) and optimizing the resulting function

$$D(\lambda) = -1 + \sum_{j=1}^{m} \lambda_j \, b_j - \ln \left\{ \sum_{i=1}^{n} q_i \, \exp\{-1 + \sum_{j=1}^{m} \lambda_j \, a_{ji}\} \right\}.\tag{11.5}$$

Since $D(\lambda)$ is continuously differentiable and concave with respect to λ, we can derive the optimal solution, λ^*, by solving

$$\nabla_\lambda D(\lambda) = 0 ,\tag{11.6}$$

which can be written in the explicit component-wise form

$$\nabla_{\lambda_j} D(\lambda) = b_j - \frac{\sum_{i=1}^{n} a_{ji} \, q_i \exp \left\{ -1 + \sum_{k=1}^{m} \lambda_j \, a_{ki} \right\}}{\sum_{i=1}^{n} q_i \exp \left\{ -1 + \sum_{k=1}^{m} \lambda_j \, a_{ki} \right\}} = 0\tag{11.7}$$

for $j = 1, \ldots, m$. The optimal vector $\lambda^* = (\lambda_1^*, \ldots, \lambda_m^*)$ can be found by solving (11.7) numerically. Finally, substitute $\lambda = \lambda^*$ and $\beta = \beta(\lambda^*)$ back into (11.2) to obtain the solution to the original MinxEnt program.

CHAPTER 12

RANDOM NUMBER, RANDOM VARIABLE, AND STOCHASTIC PROCESS GENERATION

2.1 The inverse-transform method yields the following rule for generating from the discrete uniform distribution on $\{0, 1, \ldots, n\}$. Generate $U \sim \text{U}(0,1)$ and output $\lfloor (n+1)U \rfloor$, where $\lfloor a \rfloor$ denotes the *floor* of a, that is, the largest integer smaller than or equal to a.

2.2 The pdf of the $\text{Beta}(1, \beta)$ distribution is $f(x) = \beta(1-x)^{\beta-1}$, $x \in (0,1)$. Hence, the cdf is

$$F(x) = 1 - (1-x)^\beta, \quad x \in (0,1).$$

Solving $y = 1 - (1-x)^\beta$ with respect to x gives the inverse

$$F^{-1}(y) = 1 - (1-y)^{1/\beta}.$$

Hence, to generate $X \sim \text{Beta}(1, \beta)$, draw $U \sim \text{U}(0,1)$ and return $X = 1 - U^{1/\beta}$.

2.3 The cdf of the $\text{Weib}(\alpha, \lambda)$ distribution is

$$F(x) = 1 - e^{-(\lambda x)^\alpha}, \quad x \geqslant 0.$$

Solving $y = 1 - e^{-(\lambda x)^\alpha}$ with respect to x gives the inverse

$$F^{-1}(y) = \frac{1}{\lambda}(-\ln(1-y))^{\frac{1}{\alpha}}.$$

Hence, to generate $X \sim \text{Weib}(\alpha, \lambda)$, draw $U \sim \text{U}(0,1)$ and return $X = (-\ln U)^{1/\alpha}/\lambda$.

2.4   Let $X \sim \text{Pareto}(\alpha, \lambda)$. The cdf of the Pareto distribution is

$$F(x) = \mathbb{P}(X \leqslant x) = \alpha\lambda \int_0^x (1 + \lambda u)^{-(\alpha+1)} \, du = 1 - (1 + \lambda x)^{-\alpha}, \quad x \geqslant 0,$$

where $\alpha, \lambda > 0$. It follows that the inverse of F is

$$F^{-1}(y) = \lambda^{-1} \left([1 - y]^{-1/\alpha} - 1 \right), \quad y \in (0, 1).$$

Therefore, to generate a $X \sim \text{Pareto}(\alpha, \lambda)$, generate $U \sim \text{U}(0, 1)$ and output $X = \lambda^{-1} \left(U^{-1/\alpha} - 1 \right)$. Here, we have used the fact that if $U \sim \text{U}(0, 1)$ then also $1 - U \sim \text{U}(0, 1)$.

2.5   The cdf of Y satisfies $F(y) = \mathbb{P}(Y \leqslant y) = \mathbb{P}(\mu + \sigma X \leqslant Y) = \mathbb{P}(X \leqslant (x - \mu)/\sigma) = F((x - \mu)/\sigma; 0, 1) = F(x; \mu, \sigma)$ by the definition of a location-scale family. We assume here that $\sigma > 0$.

2.6   The cdf is

$$F(x) = \begin{cases} \frac{1}{2} e^{-\lambda(\theta - x)} & x \leqslant \theta \\ 1 - \frac{1}{2} e^{-\lambda(x - \theta)} & x > \theta. \end{cases}$$

Consequently, the inverse is

$$F^{-1}(y) = \begin{cases} \theta + \frac{1}{\lambda} \ln(2y) & y \in (0, 1/2] \\ \theta - \frac{\ln(2(1-y))}{\lambda} & y \in (1/2, 1). \end{cases}$$

Thus, to generate from the Laplace distribution, first draw $U \sim \text{U}(0, 1)$. If $U \leqslant 1/2$, then output $X = \theta + \ln(2U)/\lambda$; otherwise, output $X = \theta - \ln(2U)/\lambda$. (Note again that $1 - U$ has the same distribution as U.)

2.7   The inverse of the cdf of the extreme value distribution is

$$F^{-1}(y) = \mu + \sigma \ln(-\ln(1 - y)).$$

Hence, to generate X from the extreme value distribution, draw $U \sim \text{U}(0, 1)$ and return $X = \mu + \sigma \ln(-\ln U)$.

2.8   The cdf of the triangular distribution is

$$F(x) = \begin{cases} 0 & x < 2a \\ \frac{1}{2} \left(\frac{x - 2a}{b - a} \right)^2 & 2a \leqslant x < a + b \\ 1 - \frac{1}{2} \left(\frac{2b - x}{b - a} \right)^2 & a + b \leqslant x < 2b \\ 1 & x \geqslant 2b. \end{cases}$$

Solving $\frac{1}{2} \left(\frac{x - 2a}{b - a} \right)^2 = y$ in terms of x, with $2a \leqslant x < a + b$, gives

$$x = 2a + (b - a)\sqrt{2y}, \quad (y \in (0, 1/2)),$$

and solving $1 - \frac{1}{2} \left(\frac{2b - x}{b - a} \right)^2 = y$ for $a + b \leqslant x \leqslant 2b$ gives

$$x = 2b - (b - a)\sqrt{2(1 - y)}, \quad (y \in [1/2, 1)).$$

Thus, to generate a random variable X from this triangular distribution using the inverse-transform method, first draw $U \sim \mathsf{U}(0,1)$. If $U < 1/2$, return $X = 2a + (b-a)\sqrt{2U}$; otherwise, return $X = 2b - (b-a)\sqrt{2(1-U)}$.

2.9 The cdf F is a continuous, increasing, piece-wise linear function. We assume $C_i > 0$ for all i. Define $a_i = C_i (x_i - x_{i-1})$ and $F_i = \sum_{k=1}^{i} a_k$, $i = 1, \ldots, n$. Set $F_0 = 0$. Then $F(x_i) = F_i$, $i = 0, 1 \ldots, n$. The inverse-transform method prescribes that if $U \sim \mathsf{U}(0,1) \in [F_{i-1}, F_i)$, the value $X = x_{i-1} + (U - F_{i-1})/C_i$ is returned, $i = 1, \ldots, n$.

2.10 (a) First, observe that $F_i = F(x_i)$, $i = 1, \ldots, n$. Next, for $x \in [x_{i-1}, x_i)$ we have

$$F(x) = F_{i-1} + \int_{x_{i-1}}^{x} f(u)\, du = F_{i-1} + \int_{x_{i-1}}^{x} C_i u\, du = F_{i-1} + \frac{C_i}{2}(x^2 - x_{i-1}^2).$$

(b) Set $F_0 = 0$. Solving $y = F_{i-1} + \frac{C_i}{2}(x^2 - x_{i-1}^2)$ for x gives

$$x = \sqrt{\frac{2(y - F_{i-1}) + C_i\, x_{i-1}^2}{C_i}}.$$

Hence, the inverse-transform method yields the following rule for generating from f. Generate $U \sim \mathsf{U}(0,1)$. If $U \in [F_{i-1}, F_i]$, return $X = \sqrt{\frac{2(U - F_{i-1}) + C_i\, x_{i-1}^2}{C_i}}$, $i = 1, \ldots, n$.

2.11 The cdf of the Cauchy distribution is

$$F(x) = \int_{-\infty}^{x} \frac{1}{\pi} \frac{1}{1+x^2}\, dx = \frac{1}{2} + \frac{\arctan(x)}{\pi}, \quad x \in \mathbb{R}.$$

Its inverse is

$$F^{-1}(y) = \tan(\pi y - \pi/2), \quad y \in (0,1).$$

Hence, to generate X from the Cauchy distribution, draw $U \sim \mathsf{U}(0,1)$ and return $X = \tan(\pi U - \pi/2) = -\cot(\pi U)$.

2.12 Let U and $V > 0$ be continuous random variables with joint pdf $f_{U,V}$. Letting $W = U/V$, we have

$$F_W(w) = \mathbb{P}(W \leqslant w) = \mathbb{P}(U \leqslant Vw) = \int_{v=0}^{\infty} \int_{u=0}^{vw} f_{U,V}(u,v)\, du\, dv$$

$$= \int_{v=0}^{\infty} \int_{y=0}^{w} f_{U,V}(vy, v)\, v\, dy\, dv = \int_{y=0}^{w} \int_{v=0}^{\infty} f_{U,V}(vy, v)\, v\, dy\, dv.$$

Differentiating the first and last term with respect to w gives

$$f_W(w) = \int_{v=0}^{\infty} f_{U,V}(vw, v)\, v\, dv. \qquad (12.1)$$

Now take X and Y independent and standard normal. From (12.1), the pdf of $W = X/|Y|$ is given by

$$f_W(w) = \int_{v=0}^{\infty} \frac{1}{\pi} e^{-\frac{1}{2}((wv)^2 + v^2)}\, v\, dv$$

$$= \frac{1}{\pi} \int_{v=0}^{\infty} e^{-\frac{1}{2}v^2(w^2+1)} v\, dv$$

$$= \frac{1}{\pi} \left[\frac{e^{-\frac{1}{2}v^2(w^2+1)}}{w^2+1} \right]_0^{\infty} = \frac{1}{\pi} \frac{1}{w^2+1},$$

which is the pdf of the Cauchy distribution. Finally, observe that the distribution of $X/|Y|$ is the same as that of X/Y, so the latter also has a Cauchy distribution.

2.13   If X is generated via **Algorithm 2.3.4**, then

$$\mathbb{P}(X \leqslant x) = \sum_{i=1}^{m} \mathbb{P}(X \leqslant x \mid Y = i)\, p_i = \sum_{i=1}^{m} G_i(x)\, p_i = F(x) \,.$$

Hence, X has the desired cdf F.

2.14   The composition method suggests the following algorithm.

Algorithm.

1. *Let $Y = 1, 2, 3$ with probability $1/2, 1/3$, and $1/6$, respectively.*

2. *Given $y = i$, draw X from the $N(a_i, b_i^2)$ distribution, with $(a_1, a_2, a_3) = (-1, 0, 1)$ and $(b_1, b_2, b_3) = (1/4, 1, 1/2)$.*

Here is a possible Matlab program:

```
p = [1/2,1/3,1/6];
a = [-1, 0, 1];
b = [1/4, 1, 1/2];
N = 10000;
x = zeros(1,N);
for i=1:N
   y = min(find(cumsum(p)> rand));
   x(i) = randn*b(y) + a(y);
end
hist(x,100) % make a histogram of the data
```

2.15   The function

$$w(x) = \frac{f(x)}{g(x)} = \frac{e^{-\frac{1}{2}x^2 + x}}{\sqrt{2\pi}}$$

is non-negative for $x \geqslant 0$ and is maximal for $x = 1$. The maximal value is $C = \sqrt{2e/\pi}$.

2.16   The Laplace transform of $X \sim \text{Gamma}(\alpha, 1)$ is

$$\mathbb{E}[e^{-sX}] = \left(\frac{1}{1+s}\right)^{\alpha} \,.$$

The Laplace transform of X/λ is therefore

$$\mathbb{E}[e^{-sX/\lambda}] = \left(\frac{1}{1+s/\lambda}\right)^{\alpha} = \left(\frac{\lambda}{\lambda+s}\right)^{\alpha} \,,$$

which is the Laplace transform of the $\text{Gamma}(\alpha, \lambda)$ distribution (see also **Example 1.9**).

2.17   Using the conditioning formula **(1.11)** and **Example 1.9**, the Laplace transform of $XU^{1/\alpha}$ is found to be

$$\mathbb{E}[e^{-sXU^{1/\alpha}}] = \mathbb{E}[\mathbb{E}[e^{-sXU^{1/\alpha}} \mid U]] = \mathbb{E}\left[\left(\frac{1}{1+sU^{1/\alpha}}\right)^{\alpha+1}\right]$$

$$= \int_0^1 \left(\frac{1}{1+su^{1/\alpha}}\right)^{\alpha+1} du$$

$$= \int_1^{1+s} \alpha \left(\frac{y-1}{sy}\right)^{\alpha-1} \frac{1}{sy^2} \, dy \qquad \text{(ch. of var.: } y = 1+su^{1/\alpha})$$

$$= \left[\left(\frac{y-1}{sy}\right)^\alpha\right]_1^{1+s} = \left(\frac{1}{1+s}\right)^\alpha,$$

which is the Laplace transform of the Gamma$(\alpha, 1)$ distribution.

2.18   This is an application of the transformation rule **(1.20)**. Let $\mathbf{X} = (X_1, X_2)^T$ with $X_1 \sim$ Gamma$(\alpha, 1)$ and $X_2 \sim$ Gamma$(\beta, 1)$ independent. Consider the transformation $\mathbf{x} \mapsto \mathbf{z} = (z_1, z_2)^T$ defined by

$$z_1 = \frac{x_1}{x_1 + x_2},$$

$$z_2 = x_1 + x_2.$$

The transformation is invertible, with

$$x_1 = z_1 z_2,$$

$$x_2 = (1 - z_1) z_2.$$

The corresponding Jacobian matrix is

$$\begin{pmatrix} \frac{\partial x_1}{\partial z_1} & \frac{\partial x_1}{\partial z_2} \\ \frac{\partial x_2}{\partial z_1} & \frac{\partial x_2}{\partial z_2} \end{pmatrix} = \begin{pmatrix} z_2 & z_1 \\ -z_2 & 1 - z_1 \end{pmatrix}.$$

Hence, the determinant of the Jacobian matrix is z_2. We now have

$$f_{\mathbf{Z}}(\mathbf{z}) = \frac{(z_1 z_2)^{\alpha-1} e^{-z_1 z_2}}{\Gamma(\alpha)} \frac{((1-z_1)z_2)^{\beta-1} e^{-(1-z_1)z_2}}{\Gamma(\beta)} z_2.$$

This is of the form

$$f_{\mathbf{Z}}(\mathbf{z}) = z_1^{\alpha-1}(1 - z_1)^{\beta-1} h(z_2),$$

where $h(z_2)$ depends on z_2 but not on z_1. It follows that the (marginal) distribution of Z_1 is Beta(α, β).

2.19   The acceptance-rejection procedure is similar to the $\lambda = 1$ case in **Figure 2.5**. We need to bound $f(x) = \sqrt{2/\pi} e^{-x^2/2}$ by $Cg(x) = C\lambda e^{-\lambda x}$ for some C. The ratio $f(x)/g(x)$ is maximal for $x = \lambda$, and the maximal value is $C(\lambda) = e^{\lambda^2/2}\sqrt{2/\pi}/\lambda$. The smallest value possible for $C(\lambda)$ is $C(1) = \sqrt{2e/\pi}$. The highest acceptance probability is thus obtained for $\lambda = 1$.

2.20 (a) The cdf of this truncated exponential distribution is

$$F(x) = \frac{1 - e^{-x}}{1 - e^{-a}}, \quad x \in [0, a] .$$

Its inverse is

$$F^{-1}(y) = -\ln\left(1 - (1 - e^{-a})y\right), \quad y \in (0, 1) .$$

Therefore, to generate a random variable X from this distribution using the inverse-transform method, draw $U \sim U(0, 1)$ and return $X = -\ln\left(1 - (1 - e^{-a})U\right)$.

(b) Suppose we use an $\mathsf{Exp}(\lambda)$ proposal distribution. Let

$$w(x) = \frac{f(x)}{g(x)} = \frac{e^{x(\lambda - 1)}}{\lambda(1 - e^{-a})}, \quad x \in [0, a] .$$

For $\lambda < 1$, $w(x)$ is monotonically decreasing and maximal at $x = 0$, with maximal value $C(\lambda) = 1/(\lambda(1 - e^{-a}))$. For $\lambda > 1$, $w(x)$ is monotonically increasing and maximal at $x = a$, with maximal value $C(\lambda) = e^{a\lambda}/(\lambda(e^a - 1))$. The minimal value for $C(\lambda)$ is attained at $\lambda = 1$, giving an efficiency $C(1) = 1/(1 - e^{-a})$.

(c) For $a = 1$ the (best) efficiency is $1/(1 - e^{-1}) \approx 1.582$. For $a \to 0$ the efficiency, $1/(1 - e^{-a})$, approaches 0. For $a \to \infty$ the efficiency tends to 1.

2.21 (a) The cdf is

$$F(x) = \begin{cases} 0 & x < 0 \\ \frac{1}{4}x & 0 \leqslant x < 1 \\ \frac{1}{2}x^2 - \frac{3}{4}x + \frac{1}{2} & 1 \leqslant x \leqslant 2 \\ 1 & x > 2 . \end{cases}$$

Its inverse is

$$F^{-1}(y) = \begin{cases} 4y & 0 < y \leqslant 1/4 \\ \frac{1}{4}\left(\sqrt{32y - 7} + 3\right) & 1/4 < y < 1 . \end{cases}$$

Thus, to generate from this distribution using the inverse-transform method, first draw $U \sim U(0, 1)$. If $U \leqslant 1/4$, return $X = 4U$; otherwise, return $X + \frac{1}{4}\left(\sqrt{32U - 7} + 3\right)$.

(b) Using this $U[0, 2]$ proposal distribution gives an efficiency of $C = \max_{0 \leqslant x \leqslant 2} 2 f(x) = 2 f(2) = 5/2$. The acceptance-rejection procedure to generate a random variable X from this distribution is thus as follows.

Algorithm.

1. *Generate $X \sim U[0, 2]$.*
2. *Generate $Y \sim U[0, 5/4]$ independently of X.*
3. *If $Y \leqslant f(X)$, return X; otherwise, return to Step 1.*

2.22 (a) The cdf is given by

$$F(x) = \begin{cases} 0 & x < 0 \\ \frac{x^2}{4} & 0 \leqslant x < 1 \\ \frac{1}{4} + \frac{x-1}{2} & 1 \leqslant x \leqslant 5/2 \\ 1 & x > 5/2 . \end{cases}$$

Hence, to generate a random variable X from this distribution, using the inverse-transform method, first draw $U \sim U(0, 1)$. If $\frac{1}{4} \leqslant U < 1$, output $X = \frac{1}{2} + 2U$; otherwise, output $X = 2\sqrt{U}$.

(b) The acceptance-rejection method with proposal density $g(x) = 8x/25$ has an acceptance probability of $C = \max_{0 \leqslant x \leqslant 5/2} 25f(x)/(8x) = 25f(1)/8 = 25/16$. To draw a random variable $X \sim g$, we can generate $U \sim U(0,1)$ and return $X = \frac{5}{2}\sqrt{U}$ (this is the inverse-transform way of drawing from g). Based on **Algorithm 2.3.6,** the acceptance-rejection algorithm is as follows.

Algorithm.

1. *Generate $U_1 \sim U(0,1)$ and let $X = \frac{5}{2}\sqrt{U_1}$.*
2. *Generate $U_2 \sim U(0,1)$ independently of U_1.*
3. *If $U_2 \leqslant \frac{f(X)}{Cg(X)} = \begin{cases} 1, & for\ 0 < X < 1 \\ \frac{1}{X}, & for\ 1 \leqslant X < \frac{5}{2} \end{cases}$, output $Z = X$ and exit; otherwise, repeat from Step 1.*

A histogram involving 10,000 random variables with pdf f is given in Figure 12.1.

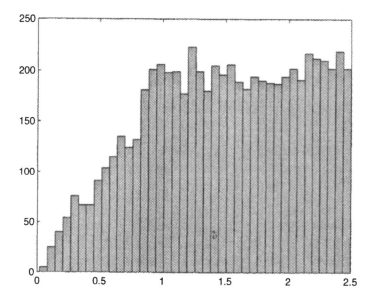

Figure 12.1 Histogram of 10,000 samples from f.

2.23 (a) First, observe that the normalization constant is $c = (1 - (1 - p)^n)^{-1}$. The cdf of this truncated geometric distribution is therefore

$$F(x) = \left\lfloor \frac{1 - (1-p)^x}{1 - (1-p)^n} \right\rfloor , \quad 1 \leqslant x \leqslant n ,$$

and its inverse is

$$F^{-1}(y) = 1 + \left\lfloor \frac{\ln\left(1 - (1 - (1-p)^n)\,y\right)}{\ln(1-p)} \right\rfloor , \quad 0 \leqslant y < 1 . \tag{12.2}$$

Thus, to generate a random variable X from this distribution, first draw $U \sim U(0,1)$ and return $X = F^{-1}(U)$ as per (12.2).

(b) The efficiency, C, is equal to $\max_x f(x)/g(x) = c = (1 - (1-p)^n)^{-1}$. For $n = \infty$, the efficiency is obviously $C = 1$, as there is no truncation. For $n = 2$, the efficiency is $C = 1/(p(2-p))$.

2.24   Let $Z_i = \max_{j=1,\ldots,r} X_{ij}$, $i = 1,\ldots,m$. Then, $Z_1,\ldots,Z_m$ are iid with cdf $F_Z(x) = [F(x)]^m$. Thus, $Y = \min_{i=1,\ldots,m} Z_i$ has cdf $F_Y(x) = 1 - (1 - F_Z(x))^m = 1 - (1 - [F(x)]^r)^m$. Solving $y = F_Y(x)$ in terms of x, we find that the inverse of F_Y — to be used in the inverse-transform method — is given by

$$F_Y^{-1}(y) = F^{-1}\left(\{1 - (1-y)^{1/m}\}^{1/r}\right).$$

2.25   A single line of Matlab code that generates one such a bargraph (see Figure 12.2) is:

```
bar((rand(1,100) < 0.2));
```

Change 0.2 to 0.5 in the code above to create a bargraph for the case $p = 0.5$.

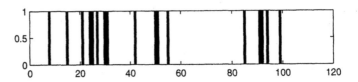

Figure 12.2 Bargraph for a Bernoulli sequence with success probability $p = 0.2$.

2.26   Let $N(t)$ denote the number of arrivals in $[0,t]$. Realizations of $\{N(t), t \in [0,1]\}$ for both the homogeneous and nonhomogeneous case are given in Figures 12.3 and 12.4, respectively.

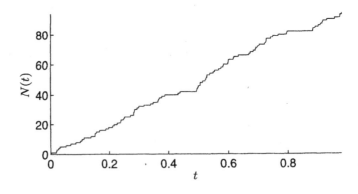

Figure 12.3 The counting process for the homogeneous Poisson process.

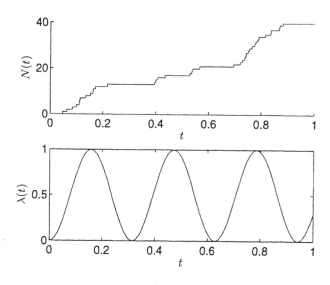

Figure 12.4 The counting process for the nonhomogeneous Poisson process with rate function $\lambda(t)$.

The following Matlab code was used to generate Figure 12.4.

```
rand('state',12345);
lambda = 100;
t1 = 0;
t = 0;
n = 0;
tt = [t];
while t <= 1
   t1 = t1 - log(rand)/lambda;
   if t1 > 1, break, end
   if (rand < sin(t1*10)^2)
       tt = [tt,t1];
       n = n+1;
   end
end
nn = 0:n;
subplot(2,1,1)
for i =1:n
   line([tt(i),tt(i+1)],[nn(i),nn(i)]);
   line([tt(i+1),tt(i+1)],[nn(i),nn(i+1)]);
end
line([tt(n+1),1],[nn(n+1),nn(n+1)]);
axis([0,1,0,n])
subplot(2,1,2)
x = 0:0.01:1;
y = sin(x*10).^2;
plot(x,y)
```

2.27   A plot of a typical realization of the two-dimensional Poisson process is given in Figure 12.5. In this case, seven points fall in the square $[1, 3]^2$. The exact distribution of the number of points in this square is Poi(4×2). Hence, the expected number of points is 8.

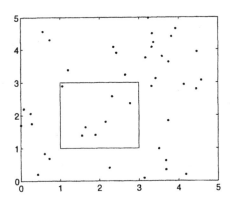

Figure 12.5   A realization of a two-dimensional homogeneous Poisson process with rate 2.

The following code was used.

```
% first, draw a Poi(50) random variable, n.
lambda = 50;
t = 0;
n = 0;
while t <= 1
    t = t - log(rand)/lambda;
    if t > 1, break, end
    n = n+1;
end
% next, draw n points uniformly distributed over [0,5]^2
x = 5*rand(1,n);
y = 5*rand(1,n)
% finally, plot the results
plot(x,y,'.');
line([1,1],[1,3]),line([1,3],[3,3]),line([3,3],[3,1])
line([3,1],[1,1])
```

2.28   Let $\mathbf{x} = (x, y)^T$. The equation for the ellipse is $\mathbf{x}^T \Sigma \mathbf{x} = 9$, with

$$\Sigma = \begin{pmatrix} 5 & \frac{21}{2} \\ \frac{21}{2} & 25 \end{pmatrix} .$$

The Cholesky decomposition of Σ is $B B^T = \Sigma$, with

$$B = \begin{pmatrix} \sqrt{5} & 0 \\ \frac{21}{2\sqrt{5}} & \sqrt{\frac{59}{20}} \end{pmatrix} . \tag{12.3}$$

Similar to **Algorithm 2.5.5** we can generate a uniform variable in the interior of the ellipse as follows.

Algorithm.

1. *Generate independently $X_1, X_2 \sim N(0,1)$ and $U \sim U[0,1]$.*

2. *Set $\mathbf{Y} = r \times \sqrt{U} \times \left(\frac{X_1}{Z}, \frac{X_2}{Z}\right)$ with $Z = \sqrt{\sum_{i=1}^{2} X_i^2}$ and $r = 3$, so that $\mathbf{Y}$ is uniformly distributed over the disc with radius r.*

3. *Output $\mathbf{X} = (B^T)^{-1}\mathbf{Y}$, with B as in (12.3).*

A typical outcome of 1,000 independent trials is given in Figure 12.6.

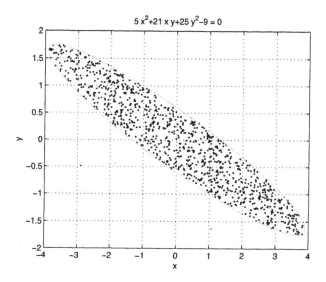

Figure 12.6   1,000 samples drawn uniformly on the interior of the ellipse $5x^2 + 21xy + 25y^2 - 9 = 0$.

2.29   The first permutation generation algorithm, **Algorithm 2.8.1**, is very easily implemented in Matlab. For example, the following code provides a uniform permutation of $(1, \ldots, 100)$.

```
[dummy,permutation] = sort(rand(1,100));
permutation
```

However, the second algorithm, **Algorithm 2.8.2**, is more difficult to implement in Matlab in an efficient way. Instead, we implemented both algorithms in C, in order to obtain a fairer comparison. We implemented the second algorithm with the shuffle modification (see **Remark 2.8.2**). Figure 12.7 plots the CPU time for obtaining 10 permutations of length

n, against n, with $n = 2^k, k = 1, \ldots, 20$. We see that the second algorithm is clearly more efficient, especially for larger n.

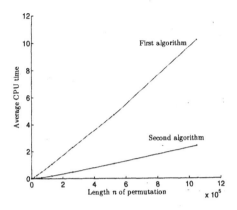

Figure 12.7 The efficiency of the two random permutation algorithms.

2.30 (a) The one-step transition matrix is

$$P = \begin{pmatrix} 0 & 1/2 & 1/2 & 0 & 0 & 0 \\ 1/3 & 0 & 1/3 & 1/3 & 0 & 0 \\ 1/4 & 1/4 & 0 & 1/4 & 1/4 & 0 \\ 0 & 1/4 & 1/4 & 0 & 1/4 & 1/4 \\ 0 & 0 & 1/3 & 1/3 & 0 & 1/3 \\ 0 & 0 & 0 & 1/2 & 1/2 & 0 \end{pmatrix}.$$

(b) Simply check that $\pi P = \pi$ and that the $\{\pi_i\}$ sum up to 1.

(c) A typical realization of the process $\{X_n, n = 0, 1, \ldots, 50\}$ starting from $X_0 = 1$ is given in Figure 12.8.

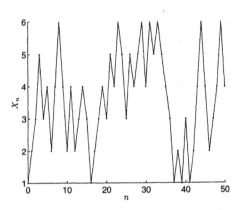

Figure 12.8 A realization of the six-state Markov chain with transition matrix P.

When keeping record of the number of visits to states $1, \ldots, 6$ over a time period of $10,000$, we found the frequencies $(0.1123, 0.1664, 0.2141, 0.2218, 0.1683, 0.1171)$, which are close to the true limiting probabilities $(0.1111, 0.1667, 0.2222, 0.2222, 0.1667, 0.1111)$.

The following Matlab code was used to make Figure 12.8.

```
clf
P =[0,1/2,1/2,0,0,0;
    1/3,0,1/3,1/3,0,0 ;
    1/4,1/4,0,1/4,1/4,0;
    0,1/4,1/4,0,1/4,1/4;
    0,0,1/3,1/3,0,1/3;
    0,0,0,1/2,1/2,0];
N = 51;
x = zeros(1,N);
x(1) = 1;
for i=2:N
    x(i) = min(find(rand < cumsum(P(x(i-1),:))));
end
t = 0:N-1
hold on
plot(t,x,'.')
plot(t,x)
hold off
```

2.31 Typical realizations for the cases $p = 1/2$ and $p = 2/3$ are given in Figure 12.9.

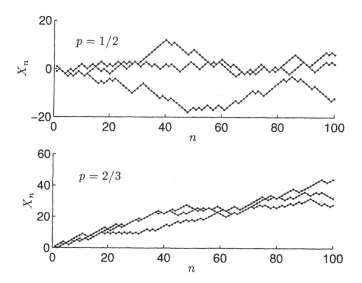

Figure 12.9 Realizations of random walks on the integers.

2.32 Two typical realizations, for $(\lambda, \mu) = (1, 2)$ and $(\lambda, \mu) = (10, 11)$, are given in Figures 12.10 and 12.11, respectively. Note that both graphs depict the number of customers in the system, not only in the queue itself.

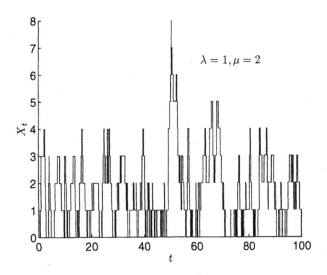

Figure 12.10 Realization of the $M/M/1$ queuelength process for $\lambda = 1, \mu = 2$.

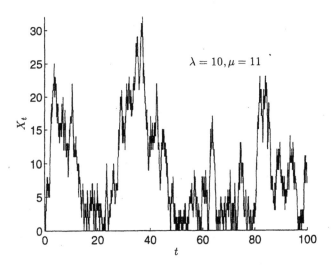

Figure 12.11 Realization of the $M/M/1$ queuelength process for $\lambda = 10, \mu = 11$.

The following Matlab code was used.

```
T = 100;
n=0;
lam = 10;
mu = 11;
t = 0;
y = 0;
yy=[y];
tt = [t];
while t < T
   if y==0
       a = -log(rand)/lam;
       y = 1;
   else
       a = -log(rand)/(lam + mu);
       y = y + 2*(rand < lam/(lam+mu)) -1;
   end
   t = t+ a;
   tt = [tt,t];
   yy = [yy,y];
   n=n+1;
end
for i =1:n
   line([tt(i),tt(i+1)],[yy(i),yy(i)]);
   line([tt(i+1),tt(i+1)],[yy(i),yy(i+1)]);
end
axis([0,T,0,max(yy)]);
```

CHAPTER 13

SIMULATION OF DISCRETE-EVENT SYSTEMS

3.1   The following Matlab program implements an event-oriented simulation implementation
of the $M/M/1$ queueing system. The structure is very similar to that for the more complex
tandem queue in **Figures 3.6** and **3.7**.

```
%main.m
clear all;
mu = 3; lambda = 2;
rho = lambda/mu;
T = 5000;
n1 = 0;
nn1=n1;
ev_list = inf*ones(2,2);
t = 0;
tot = 0;
tt = 0;
ev_list(1,:) = [- log(rand)/lambda, 1]; %schedule the first arrival
N_ev = 1;                % number of scheduled events
while t < T
  t = ev_list(1,1);
  tt=[tt,t];
  ev_type = ev_list(1,2);
  switch ev_type
   case 1
    arrival
```

```
  case 2
    departure
  end
  N_ev = N_ev - 1;
  ev_list(1,:) = [inf,inf];
  ev_list = sortrows(ev_list,1); % sort event list
nn1=[nn1,n1];
tot =tot + nn1(end-1)*(tt(end) - tt(end-1));
end
res = tot/t
exact = rho/(1-rho)
```

```
%arrival.m
N_ev = N_ev + 1;
ev_list(N_ev,:) = [t - log(rand)/lambda, 1];  %schedule new arrival
if n1 == 0  % if queue is empty
  N_ev = N_ev + 1;
  ev_list(N_ev,:) = [t - log(rand)/mu, 2]; % schedule  departure at queue 1
end
n1 = n1+1;
```

```
%departure.m
n1 = n1-1; % go out queue
if n1 ~= 0
  N_ev = N_ev + 1;
  ev_list(N_ev,:) = [t - log(rand)/mu, 2]; % schedule  departure
                                           % at queue 2
end
```

Figure 13.1 depicts a sample path of the $M/M/1$ queue with arrival rate $\lambda = 1$ and service rate $\mu = 2$. Figure 13.2 depicts a sample path of the $M/M/1$ queue with arrival rate $\lambda = 10$ and service rate $\mu = 11$. See also Problem 2.32, where the *same* system is simulated in a different way.

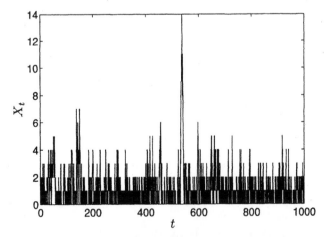

Figure 13.1 Realization of the $M/M/1$ queue with $\lambda = 1$ and $\mu = 2$.

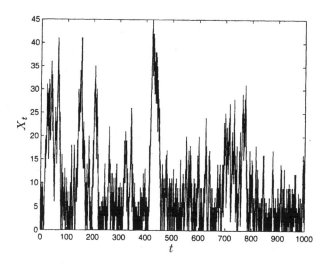

Figure 13.2 Realization of the $M/M/1$ queue with $\lambda = 10$ and $\mu = 11$.

3.2 Figure 13.3 depicts a sample path of a queue with $U(0, 2)$-distributed interarrival times and $U(0, 1/2)$-distributed service times. The same program as in Problem 3.1 can be used, with the exponential service and interarrival times replaced with uniform ones.

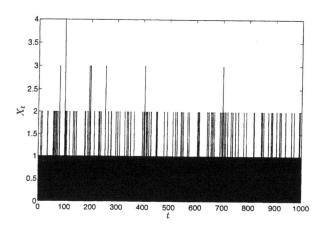

Figure 13.3 Realization of a $GI/G/1$ queue with mutually independent $U(0, 2)$ interarrival times and $U(0, 1/2)$ service times.

3.3 Out of 1,000 independent simulation runs, 312 led to a negative balance during the first 100 days.

For large t, we would expect the process to be near the average trajectory, which is 16 every 8.5 days, -5 every day, and 100 every 30 days, starting at 150. This yields

$$X(t) \approx 150 + ((16/8.5) + 100/30 - 5)\, t \approx 150 + 0.215686\, t\,.$$

However, this approximation is only valid for very large t, say $t > 100{,}000$ days (274 years), which renders it useless for practical predictions. To illustrate, Figure 13.4 depicts a realization over 10,000 days, and shows that the actual process can deviate significantly from the linear trend.

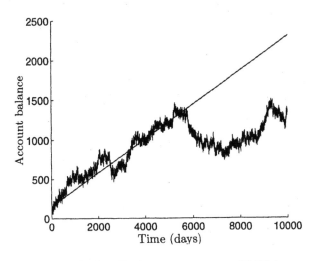

Figure 13.4   Realization of bank account process over 10,000 days.

3.4   Figure 13.5 depicts the number of customers in both queues for a realization of the tandem queue network.

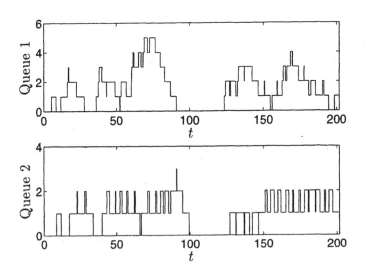

Figure 13.5   Realization of the tandem queue.

3.5   (a) Consider the two-machine one-repairman problem with exponential life and repair times. Let $X_t \in \{0, 1, 2\}$ be the number of failed machines at time t. If at time t the process is in state 1, it will remain there until the failed machine is repaired or the working machine fails. However, because of the *memoryless property* of the exponential distribution, knowledge of how long the repair has been going on before time t or of the age of the working machine at time t is irrelevant. The same reasoning can be applied to the other states. More precisely, the stochastic process $X = \{X_t, t \geq 0\}$ satisfies the *Markov property* (**1.40**). Since the Markov process X can only jump one state up or down from the current state, it is a birth-and-death process. The birth and death rates are $b_0 = \frac{2}{5}$, $b_1 = \frac{1}{5}$, $d_1 = 1$, and $d_2 = 1$. For example, when the process is in state 0, the probability of a machine failure in the next h time units is approximately $h\,(1/5 + 1/5)$ for small h. Thus, $b_0 = 2/5$.

(b) Starting the process in state 0 (both machines working), we run the simulation ten times, from $t = 0$ to $t = T = 100{,}000$. The ten estimates of the fraction of time that both machines were out of order are given in Table 13.1 below.

Table 13.1   Ten independent estimates for the proportion of time both machines are down.

k	$\widehat{\ell}_k$	k	$\widehat{\ell}_k$
1	0.0549787	6	0.0530450
2	0.0542249	7	0.0547214
3	0.0527266	8	0.0530919
4	0.0553471	9	0.0550757
5	0.0540670	10	0.0533335

This gives an overall average estimate of $\widehat{\ell} \approx 0.054$ and an estimated relative error of approximately 0.0177, based on these ten estimates. The stationary distribution of X is $(25, 10, 2)/37$. Hence, the limiting probability of having two machines down is $2/37 \approx 0.054$. The following code was used.

```
Q = [ -2/5 2/5  0  ;
       1  -6/5 1/5 ;
       0    1   -1 ];
K = Q - diag(diag(Q));
for i=1:size(K,1)
   K(i,:)=K(i,:)./sum(K(i,:));
end

t=0; told=0;
x=1; xold=1; % both start working ...
tbo=0; % total time both machines are o
T=100000;

while t<T
   lambda = -Q(x,x);
   t=t-log(rand(1))/lambda;
   x=min(find(rand < cumsum((K(x,:)))));
   if xold==3, tbo=tbo+t-told; end
   told=t;
   xold=x;
end
if xold==3, tbo=tbo+T-told; end

fprintf('Both machines were out %g fraction of time.\n',tbo/T)
```

(c) The typical output of this process is depicted in Figure 13.6.

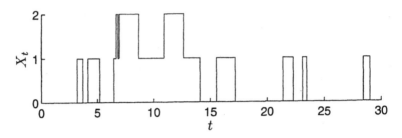

Figure 13.6   Realization of the two-machine, one-repairman process.

The following event-oriented Matlab code was used.

```
T = 30;
mach_num = 0; repairq = [];
nrep = 1; nmach = 2; % number of repairmen and machines
r = nrep; f = 0; % the number of repairmen available and
                 % the number of machines failed
rr= r; ff=f; % the history
tt=0;
ev_list = inf*ones(10,3); % record time, type, number
t = 0;
for i=1:nmach
    ev_list(i,:) = [-5*log(rand(1)), 1,i]; % schedule the failure times
end
ev_list = sortrows(ev_list,1); % sort event list
N_ev = nmach;
while t < T
    t = ev_list(1,1);
    tt=[tt,t];
    ev_type = ev_list(1,2);
    mach_num = ev_list(1,3);
    switch ev_type
        case 1  % machine fails
            N_ev = N_ev + 1;
            if (r > 0) % repair man available
               % schedule repair
                   ev_list(N_ev,:) = [t-log(rand(1)), 2,mach_num];
                   r = r -1;
            else
                   repairq = [repairq,mach_num];
            end
            f = f+ 1;
        case 2  % machine is repaired
            f = f - 1; % one less failed
            sq = size(repairq,2);
            if (sq > 0) % still one in the queue
                N_ev = N_ev + 1;
                % schedule next repair
                ev_list(N_ev,:) = [t-log(rand(1)), 2,repairq(1)];
                repairq = repairq(2:sq); % remove machine
```

```
                else
                    r = r+1;
                end
                N_ev = N_ev + 1;
                ev_list(N_ev,:) = [t-5*log(rand(1)), 1,mach_num]; % schedule
                                                  %failure of current machine
            end
            N_ev = N_ev - 1;
            ev_list(1,:) = [inf,inf,inf];
            ev_list = sortrows(ev_list,1); % sort event list

            rr=[rr,r];ff=[ff,f];
            disp([r,f]);
        end
        % plotting
        figure(1),clf
        for i =1:length(ff)-1,
            line([tt(i),tt(i+1)],[ff(i),ff(i)]);
            line([tt(i+1),tt(i+1)],[ff(i),ff(i+1)]);
        end
        xlabel('$t$')
        ylabel('$X_t$')
```

(d) Reusing the code from part (c) (replacing only the inter-failure and repair times with the required uniform ones), we run ten independent replications of the two-machine, one-repairman process from time $t = 0$ to $t = T = 100,000$. For each replication, we calculate the proportion of the total time that both machines were down. The results are tabulated in Table 13.2 below.

Table 13.2 Ten independent estimates for the proportion of time both machines are down, with $U[0, 10]$ lifetimes and $U[0, 2]$ service times.

k	$\widehat{\ell}_k$	k	$\widehat{\ell}_k$
1	0.0374900	6	0.0368734
2	0.0386765	7	0.0386524
3	0.0375525	8	0.0384854
4	0.0389319	9	0.0373812
5	0.0378231	10	0.0381267

This gives an overall average estimate of $\widehat{\ell} \approx 0.038$ and an estimated relative error of approximately 0.0179, based on these ten estimates. Note that the average time that both machines are down is lower than in the case of exponentially distributed life and service times, despite equivalence of the average life and service times.

(e) The typical output of this process (number of failed machines) is depicted in Figure 13.7. Here we used the same exponential life and repair times as in (c), making minimal adaptations to the event-oriented program.

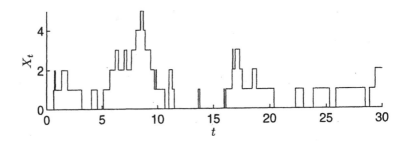

Figure 13.7 Realization of the five-machine, three-repairman process.

3.6 In Figure 13.8, the process is started with setting $n_1 = n_2 = 0$ and starting the Customer Generation process.

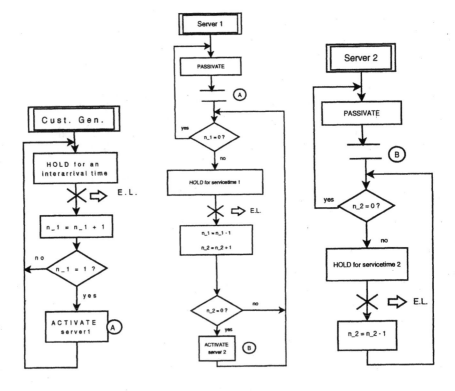

Figure 13.8 Flowcharts for the tandem queue process.

3.7 Figures 13.9–13.11 show the differences in behavior when changing the cycle time α for the continuous polling system with a greedy server. As the server's speed, $1/\alpha$, increases, the process becomes more like an $M/M/1$ queue, with the same arrival and service rates ($\lambda = 1$, $\mu = 2$).

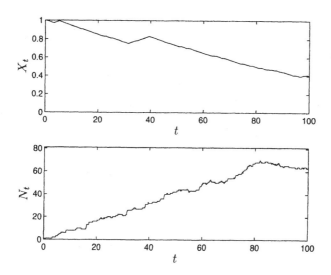

Figure 13.9 Realization of the continuous polling system with greedy server, $\alpha = 100$.

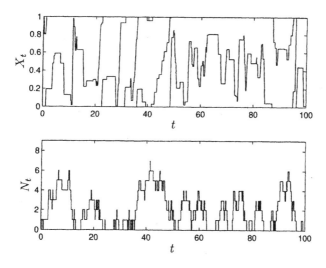

Figure 13.10 Realization of the continuous polling system with greedy server, $\alpha = 1$.

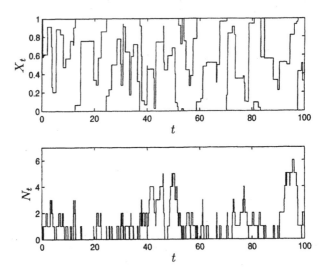

Figure 13.11 Realization of the continuous polling system with greedy server, $\alpha = 1/100$.

3.8   In Figure 13.12 the estimated throughput as a function of the buffer size b is depicted for three different simulation runs. Each run has a simulation time of 10,000. Note that for $b = 0$ the throughput is equal to the long-run average amount of time that *all* machines are working, that is, $(2/3)^3 \approx 0.3$. Similarly, for $b = \infty$ the throughput is equal to the probability that machine 1 is working, that is, $2/3 \approx 0.67$. Note that the buffer size need not be much more than 20 in order to achieve near-optimal throughput.

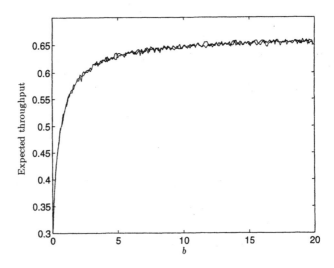

Figure 13.12 Estimated average throughput as a function of the buffer size b.

CHAPTER 14

STATISTICAL ANALYSIS OF DISCRETE-EVENT SYSTEMS

4.1. (a) The exact solution is $\ell \approx 2.3926$. Because of the variability of the estimates (see (b)), typical solutions for Methods 1 and 2 are around $2.2 - 2.6$. A simple Matlab program for this problem is as follows:

```
N = 100;
X = 4*rand(1,N)-2;
% comment one of the following:
Z = 4*exp(-X.^2/2);   % method 1
%Z = sqrt(2*pi)*(abs(X)< 2); % method 2
ell = mean(Z)
sig = std(Z)
RE = std(Z)/sqrt(N)/ell
CI = ell*[1 - 1.96*RE  1 + 1.96*RE]
```

In the calculations below we use the estimates 2.26 and 2.41, respectively.

(b) The estimated relative error for Method 1 is significantly larger than for Method 2. Typical estimates are 0.05 for Method 1 and 0.02 for Method 2.

(c) Typical 95% confidence intervals are $(2.04, 2.48)$ for Method 1 and $(2.31, 2.50)$ for Method 2. Note that the width of the latter is much smaller.

(d) The relative width of the confidence interval for Method 1 with $N = 100$ is $(2.48 - 2.04)/2.26 \approx 0.19$. To make this relative width less than 0.001, we need $(0.19/0.001)^2 \approx 38{,}000$ times as many samples; that is, N should be more than 3,800,000. Much fewer samples are required for Method 2. In this case,

$$N > 100 \left(\frac{2.50 - 2.31}{2.41 \times 0.001} \right)^2 \approx 620{,}000 \, ,$$

which roughly corresponds to a reduction with a factor of $(0.02/0.05)^2 = 0.16$ compared with Method 1.

4.2   The system works if and only if at least one of the minimal paths (1,4), (1,3,5), (2,5), or (2,3,4) works.

4.3   The reliability of the system is

$$
\begin{aligned}
\mathbb{E}[H(\mathbf{X})] &= \mathbb{E}[1 - (1 - X_1 X_4)(1 - X_2 X_5)(1 - X_1 X_3 X_5)(1 - X_2 X_3 X_4)] \\
&= \mathbb{E}[X_1 X_4 + X_2 X_3 X_4 - X_1 X_2 X_3 X_4 + X_2 X_5 + X_1 X_3 X_5 \\
&\quad - X_1 X_2 X_3 X_5 - X_1 X_2 X_4 X_5 - X_1 X_3 X_4 X_5 - X_2 X_3 X_4 X_5 \\
&\quad + 2 X_1 X_2 X_3 X_4 X_5] \, ,
\end{aligned}
$$

where we use the fact that $X_i^2 = X_i$. Since all X_i are independent and $\mathbb{E}[X_i] = p$, the last expectation becomes

$$p^2 + p^3 - p^4 + p^2 + p^3 - p^4 - p^4 - p^4 - p^4 + 2p^5 = p^2(2 + 2p - 5p^2 + 2p^3) \, .$$

4.4   The true reliability is 0.46. Typical values for a sample of size 1,000 are $\widehat{\ell} = 0.48$ with a relative error of 0.033. Thus, to obtain a relative error of 0.01, the sample size must be taken around $(0.033/0.01)^2 \cdot 1000 \approx 1.1 \cdot 10^4$. A typical result with this sample size is $\widehat{\ell} = 0.46$ with a relative error of 0.01 and a 95% confidence interval of $(0.449, 0.467)$.

4.5 (a) A typical result is $\widehat{\ell} = 5.00$ with a relative error of 0.02 and a 95% confidence interval of $(4.81, 5.19)$.

(b) A typical result is $\widehat{\ell} = 0.14$ with a relative error of 0.08 and a 95% confidence interval of $(0.11, 0.16)$.

4.6 (b) The four possible paths are $1 \rightarrow 3 \rightarrow 5 \rightarrow 6$, $1 \rightarrow 3 \rightarrow 4$, $2 \rightarrow 4$, and $2 \rightarrow 5 \rightarrow 6$.

(c) For $N = 1{,}000$ we obtained a point estimate $\widehat{\ell} = 7.2$ with an estimated relative error of 0.016 and a 95% confidence interval of $(6.9, 7.4)$. We used the following code.

```
% main.m
N = 10^3;
u = [1.1 2.3 1.5 2.9 0.7 1.5];
x = -log(rand(N,6))*diag(u);
h   = H(x);
ell = mean(h)
sig = std(h)
RE = sig/sqrt(N)/ell
CI = ell*[1 - 1.96*RE  1 + 1.96*RE]
```

Here the function H is defined as follows:

```
% H.m
function y = H(x)
y = max([x(:,1)+x(:,3)+x(:,5)+x(:,6), x(:,2)+x(:,4), ...
x(:,2)+x(:,5)+x(:,6), x(:,1)+x(:,3)+x(:,4)], [],2) ;
```

4.7   A typical outcome is a point estimate of 0.57 with an estimated 95% confidence interval of $(0.53, 0.60)$. The relative error is about 0.03. The true probability is approximately 0.58.

4.8 (a) A typical point estimate is 1.01 with an estimated 95% confidence interval of $(0.94, 1.07)$. The relative error is around 0.03. Note that the exact ℓ is equal to $\varrho/(1 - \varrho) = 1$, with $\varrho = \lambda/\mu$.

(b) The regenerative method yields (must yield!) a similar point estimate, confidence interval, and relative error. The number of regeneration cycles is around 5,000.

(c) For both methods, using $T = 10,000$ gives a relative width of around $1.96 \times 2 \times \mathrm{RE} \approx 0.12$. To obtain a width of 5% one needs a factor of $(0.12/0.05)^2 \approx 5.8$ more simulation time. Hence, $T = 60,000$ should be enough.

4.9 (a) From **Example 4.3** we know that Z — the steady-state number of people in the system as seen by an arriving customer — has distribution given by

$$\mathbb{P}(Z = k) = (1 - \varrho)\varrho^k, \quad k = 0, 1, 2, \dots ,$$

where $\varrho = \frac{\lambda}{\mu} = 0.6$. In other words, $Z + 1 \sim \mathrm{G}(1 - \varrho)$. It follows that

$$\mathbb{E}[Z] = \frac{1}{1 - \varrho} - 1 = \frac{\varrho}{1 - \varrho} = \frac{3}{2}.$$

(b) Notice that $\{Z_n, n = 1, 2, \dots\}$ is the *jump chain* of the birth-and-death process $\{X_t, t \geqslant 0\}$ on $\mathbb{N}$ with birth rates $b_i = \lambda$, $i = 0, 1, 2, \dots$ and death rates $d_i = \mu$, $i = 1, 2, \dots$. From **Example 1.12** we see that the one-step transition matrix K of the Markov chain $\{Z_n\}$ is given by

$$K(0, 1) = 1 ,$$
$$K(i, i + 1) = \frac{\lambda}{\lambda + \mu} ,$$
$$K(i, i - 1) = \frac{\mu}{\lambda + \mu}, \quad i = 1, 2, 3, \dots .$$

Hence, $\{Z_n\}$ is a random walk on $\mathbb{N}$ with $p = \frac{\lambda}{\lambda+\mu} = 0.375$ and $q = \frac{\mu}{\lambda+\mu} = 0.625$, with p and $q = 1 - p$ as in **Example 1.11** (see also Problem 4.7).

(c) A typical result is $\hat{\ell} = 1.63$ with an estimated relative error of 0.07 and a 95% confidence interval of $(1.41, 1.86)$. Because of the relatively small sample size, the relative error is not so accurately estimated. The true relative error is closer to 0.1, as can be observed by repeating the experiment a number of times and taking the sample standard deviation of the estimates. We used the following Matlab code.

```
clear all
lambda = 0.6; mu=1;
p=lambda/(lambda + mu);
q=1-p;
M=10000;
K = 100; %throw away
```

```
B = 300; %number of batches
N = M-K; %remaining samples
T = (M-K)/B;
z = zeros(1,M);
for i=2:M
    if rand<p
        z(i)=z(i-1)+1;
    elseif z(i-1) %if z is not zero
        z(i)=z(i-1)-1;
    end
end
y=zeros(1,B); %the batch means
for k=1:B
    y(k) = mean(z(101 + (k-1)*T:100 + k*T));
end
ell = mean(y);
RE = std(y)/ell/sqrt(B)
fprintf('ell %g ; 0.95 CI ( %g , %g ) \n',ell,...
            ell*(1-1.96*RE),ell*(1+1.96*RE))
```

(d) The regenerative method gives similar results to the batch means method, for example, $\hat{\ell} = 1.56$ with relative error 0.10 and a 95% confidence interval of $(1.22, 1.89)$. Note that the relative error seems larger than the one in the previous method. This is mainly due to underestimation in the first method. For large sample sizes both methods should produce similar relative errors. The example Matlab code is as follows.

```
clear all
lambda = 0.6; mu=1;
p=lambda/(lambda + mu);
q=1-p;

N=10000;
z = zeros(1,N);
R = zeros(1,N);
tau = zeros(1,N);
Rsum=0;
regcount = 0;
lastregtime = 1;
for i=2:N
    if rand<p
        z(i)=z(i-1)+1;
    elseif z(i-1) %if z is not zero
        z(i)=z(i-1)-1;
    end
    Rsum = Rsum + z(i);
    if z(i)==0 %regeneration detected
        regcount = regcount + 1;
        R(regcount) = Rsum;
        tau(regcount) = i - lastregtime;
        Rsum = 0;
        lastregtime = i;
    end
```

```
end

ell = mean(R)/mean(tau)
C = cov(R,tau); % the covariance matrix
s = sqrt(C(1,1) - 2*ell*C(1,2) + ell^2*C(2,2))
RE = s/mean(tau)/sqrt(N)
fprintf('ell %g ; 0.95 CI ( %g , %g ) \n',ell,...
         ell*(1-1.96*RE),ell*(1+1.96*RE))
```

(e) We assume that for both methods the relative error is about 0.1 for $N = 10,000$. The absolute width of the 95% confidence interval is therefore $\hat{\ell} \times \text{RE} \times 1.96 \times 2 \approx 0.588$. To obtain an absolute width of 0.05 we therefore need a factor of $(0.588/0.05)^2 \approx 140$ more samples, that is, $N = 1.4 \cdot 10^6$.

(f) Both the batch means and the regenerative methods give a confidence interval centered at 4. The confidence interval, however, is too wide and we need a larger sample size to obtain an absolute width w_a of about 5%. The absolute width w_a becomes larger as we increase the traffic intensity. It also takes a longer and longer time between regeneration cycles as $\varrho \to 1$. This is because π_0 becomes very small as the traffic intensity increases. When $\varrho = 1$, the random walk on the positive integers in part (b) becomes a null-recurrent Markov chain.

4.10　For this problem, each cycle starts when the chain enters state 0. The cycles are given by
$0 \to 3 \to 0, 0 \to 1 \to 2 \to 1 \to 0, 0 \to 2 \to 0, 0 \to 1 \to 0, 0 \to 1 \to 0$, and $0 \to 2 \to 0$.
These have cycle lengths of $\tau_1 = 2, \tau_2 = 4, \tau_3 = 2, \tau_4 = 2, \tau_5 = 2$, and $\tau_6 = 2$. The cycle rewards are given by $R_1 = 3, R_2 = 4, R_3 = 2, R_4 = 1, R_5 = 1$, and $R_6 = 2$. Hence,

$$\hat{\tau} = \frac{2+4+2+2+2+2}{6} = \frac{7}{3},$$
$$\hat{R} = \frac{3+4+2+1+1+2}{6} = \frac{13}{6},$$

which gives a point estimate of $\hat{\ell} = \hat{R}/\hat{\tau} = 13/14$.

　　The quantities of interest needed to calculate the corresponding 95% confidence interval are

$$S_{11} = \frac{1}{5}\left(\frac{25+121+1+49+49+1}{36}\right) = \frac{41}{30},$$

$$S_{22} = \frac{1}{5}\left(\frac{1+25+1+1+1+1}{9}\right) = \frac{2}{3},$$

$$S_{12} = \frac{1}{5}\left(\frac{(5)(-2)+(11)(10)+(-1)(-2)+(-7)(-2)+(-7)(-2)+(-1)(-2)}{36}\right)$$
$$= \frac{1}{5}\left(\frac{-10+110+2+14+14+2}{36}\right) = \frac{11}{15}.$$

Therefore,

$$S^2 = S_{11} - 2\hat{\ell}S_{12} + (\hat{\ell})^2 S_{22} = \frac{41}{30} - \frac{13}{7} \times \frac{11}{15} + \frac{169}{196} \times \frac{2}{3} = \frac{142}{245}.$$

This gives a 95% confidence interval of

$$\left(\hat{\ell} \pm \frac{1.96\,S}{\hat{\tau}\sqrt{6}}\right) = \left(\frac{13}{14} \pm \frac{1.96\sqrt{142/245}}{14/\sqrt{6}}\right) \approx (0.93 \pm 0.26).$$

4.11 (a) The Lindley equation is given by

$$W_{n+1} = \max\{W_n + S_n - A_{n+1}, 0\}, \quad n = 1, 2, \ldots,$$

where W_k is the waiting time of the k-th customer, S_k is the service time of the k-th customer, and A_{k+1} is the time between the k-th and $(k + 1)$-st arrivals in a $GI/G/1$ queue.

This can be explained as follows. If customer n is served before customer $n+1$ arrives (that is, $W_n + S_n < A_{n+1}$), then the server will be ready to process customer $n + 1$ immediately upon arrival and so the time that customer $n + 1$ must wait in that case is zero. On the other hand, if customer $n + 1$ arrives before customer n has finished waiting and being served, then customer $n + 1$ must wait for the remaining time that it takes for customer n to finish waiting and being served. This time is exactly $W_n + S_n$ less than the time it takes customer $n + 1$ to arrive, which is A_{n+1}.

(b) Typical output is a point estimate of $\widehat{\ell} = 0.315$ with an estimated relative error of 0.025 and a 95% confidence interval of $(0.30, 0.33)$. Example code follows.

```
lamS = 2;
lamA = 1;
N = 5000;
ww=zeros(1,N);
for i = 1:N
    w = 0; % start from an empty system
    aa=-log(rand(1,3))/lamA;
    ss=-log(rand(1,3))/lamS;
    for k=1:3
        w=max(w+ss(k)-aa(k),0);
    end
    ww(i) = w; % waiting time of 4-th customer
end
ell=mean(ww);
RE = std(ww)/ell/sqrt(N)
pm=1.96*ell*RE;
fprintf('Est: %g , CI ( %g , %g ) \n',ell,ell-pm,ell+pm)
```

(c) Typical output is a point estimate of $\widehat{\ell} = 0.50$ with an estimated relative error of 0.01 and an approximate 95% confidence interval of $(0.49, 0.51)$. Note that the exact expected steady-state waiting time of a customer is $1/(\mu - \lambda) = 1/(2 - 1) = 1/2$, where μ and λ are the service and arrival rates.

4.12   Typical simulation output is $\widehat{\ell} = 0.42$ with an estimated relative error of 0.025 and a 95% confidence interval of $(0.40, 0.44)$.

4.13   A typical estimate is $\widehat{\ell} = 2.21$ with an estimated relative error of 0.01 and a 95% confidence interval of $(2.17, 2.25)$.

4.14　The process $\{Y_i,\ i = 0, 1, \ldots\}$ is regenerative since it is a Markov chain. The state space is $\{(1,1), (1,2), (2,1), (2,2)\}$. The one-step transition matrix is

$$R = \begin{pmatrix} \frac{1}{3} & \frac{2}{3} & 0 & 0 \\ 0 & 0 & \frac{1}{4} & \frac{3}{4} \\ \frac{1}{3} & \frac{2}{3} & 0 & 0 \\ 0 & 0 & \frac{1}{4} & \frac{3}{4} \end{pmatrix},$$

where the rows and columns are ordered as $(1,1), (1,2), (2,1), (2,2)$.

We can find the limiting distribution by solving $\pi R = \pi$, where π is a row vector with nonnegative entries summing to 1. This yields the limiting distribution (by definition the distribution of the steady-state random variable Y)

$$\pi = \left(\frac{1}{11}, \frac{2}{11}, \frac{2}{11}, \frac{6}{11} \right).$$

It follows that ℓ is given by

$$\ell = \mathbb{E}[H(Y)] = \frac{c_{11}}{11} + \frac{2\,c_{12}}{11} + \frac{2\,c_{2,1}}{11} + \frac{6\,c_{2,2}}{11} = \frac{0}{11} + \frac{2}{11} + \frac{4}{11} + \frac{18}{11} = \frac{24}{11}.$$

Hence, the exact value is $\ell = 24/11 \approx 2.1818$, which is contained in the typical 95% confidence interval given in Problem 4.13.

4.15　If X_t and Y_t denote the number of customers in the first and second queue of a tandem queue at time t (including those currently being served), then the process $\{(X_t, Y_t), t \geqslant 0\}$ is a regenerative process. The process regenerates at times when a customer arrives to an empty system (both queues are empty).

4.16 (a)　A typical Monte Carlo estimate using 1,000 regenerative cycles is $\widehat{\ell} = 0.025$ with an estimated relative error of 0.1 and an estimated 95% confidence interval (based on the central limit theorem) of $(0.02, 0.03)$.

(b)　The bias is small, for example, $6.4 \cdot 10^{-5}$ and the mean square error (MSE) is around $7 \cdot 10^{-6}$. However, the bootstrapped relative error, $\sqrt{\mathrm{MSE}}/\widehat{\ell}$, which is possibly a better measure of accuracy, is not small. It is around 0.1, similar to the relative error in (a).

(c)　The 95% bootstrap confidence intervals for ℓ using the normal and percentile methods with $B = 300$ bootstrap samples, each of size 1,000, are very similar to the ones in (a), because the bootstrapped sample $\{\ell_i^*\}$ is close to being normally distributed. We used the following code.

```
clear all
mach_num = 0; repairq = [];
% number of repairmen and machines
nrep = 2; nmach = 3;
% the number of repairmen available and the number of machines failed
r = nrep; f = 0;
ff=f; % the history
tt=0;
tlast = 0; % time of last regeneration
ev_list = inf*ones(10,3); % record time, type, number
t = 0;
```

```
for i=1:nmach
    % schedule the failures (exp life time)
    ev_list(i,:) = [-10*log(rand(1)), 1,i];
end
ev_list = sortrows(ev_list,1); % sort event list
N_ev = nmach;
Nreg = 1001; % number of regenerations  + 1
regcount = 1;
R = zeros(1,Nreg);
tau = zeros(1,Nreg);
while regcount < Nreg
    t = ev_list(1,1);
    ev_type = ev_list(1,2);
    mach_num = ev_list(1,3);
    switch ev_type
        case 1  % machine fails
            N_ev = N_ev + 1;
            if (r > 0) % repair man available
                %schedule repair (uniform)
                ev_list(N_ev,:) = [t+rand*8, 2,mach_num];
                r = r -1;
            else
                repairq = [repairq,mach_num];
            end
            f = f+ 1;
        case 2  % machine is repaired
            f = f - 1; % one less failed
            sq = size(repairq,2);
            if (sq > 0) % still one in the queue
                N_ev = N_ev + 1;
                % schedule next repair
                ev_list(N_ev,:) = [t+rand*8, 2,repairq(1)];
                repairq = repairq(2:sq); % remove machine
            else
                r = r+1;
            end
            N_ev = N_ev + 1;
            % schedule failure of current machine
            ev_list(N_ev,:) = [t-10*log(rand(1)), 1,mach_num];
    end
    N_ev = N_ev - 1;
    ev_list(1,:) = [inf,inf,inf];
    ev_list = sortrows(ev_list,1); % sort event list

    if f==0 % regeneration
        tau(regcount) = t - tlast;
        regcount = regcount + 1;
        tlast = t;
    elseif f==3
        R(regcount)=R(regcount)+(t - tt(end));
    end
    tt=[tt,t];ff=[ff,f];
end
```

```
N = Nreg-1; % discarding last
R = R(1:N);
tau = tau(1:N); %discarding last
ell = mean(R)/mean(tau)
C = cov(R,tau);
s = sqrt(C(1,1) - 2*ell*C(1,2) + ell^2*C(2,2));
RE = s/mean(tau)/sqrt(N)/ell
fprintf('ell %g ; 0.95 CI ( %g , %g ) \n',ell,...
        ell*(1-1.96*RE),ell*(1+1.96*RE))

B = 300;
for b=1:B
   % resample 1000 times:
   ind = ceil(N*rand(1,N));
   Rs = R(ind);
   taus = tau(ind);
   ell_B(b) = mean(Rs)/mean(taus);
end
 bias= mean(ell_B) - ell;
 mse = sum((ell_B - ell).^2)/B;

 q_l = quantile(ell_B, 0.025);
 q_u = quantile(ell_B,0.975);
 %CI using percentile method
 fprintf('bias= %g; MSE = %g ; 0.95 CI ( %g , %g ) \n',bias,mse, q_l, q_u)
```

CHAPTER 15

CONTROLLING THE VARIANCE

5.1 For the antithetic random variable estimator $\widehat{\ell}^{(a)}$ in **(5.9)**, we have that

$$\mathrm{Var}(\widehat{\ell}^{(a)}) = \mathrm{Var}\left(\frac{1}{2N}\sum_{i=1}^{N}\{H(X_i) + H(b+a-X_i)\}\right)$$

$$= \frac{1}{4N}\,\mathrm{Var}\left(H(X_1) + H(b+a-X_1)\right)$$

$$= \frac{1}{4N}\left(\mathrm{Var}\left(H(X_1)\right) + \mathrm{Var}\left(H(b+a-X_1)\right)\right) + 2\,\mathrm{Cov}\left(H(X_1), H(b+a-X_1)\right))$$

$$= \frac{1}{4N}\left(2\,\mathrm{Var}\left(H(X_1)\right) + 2\,\mathrm{Cov}\left(H(X_1), H(b+a-X_1)\right)\right)$$

$$= \frac{1}{2N}\left(\mathrm{Var}\left(H(X_1)\right) + \mathrm{Cov}\left(H(X_1), H(b+a-X_1)\right)\right).$$

Similarly, the CMC estimator **(5.8)** has

$$\mathrm{Var}(\widehat{\ell}) = \frac{1}{N}\mathrm{Var}\left(H(X_1)\right).$$

If $H(x)$ is monotonic in x, then $\mathrm{Cov}\left(H(X_1), H(b+a-X_1)\right) \leqslant 0$, since $H(x)$ and $H(b+a-x)$ are monotonic in opposite directions. Therefore,

$$\mathrm{Var}(\widehat{\ell}^{(a)}) \leqslant \frac{1}{2}\mathrm{Var}(\widehat{\ell}).$$

5.2   Typical simulation output gives an estimate of 1.4989 with an estimated relative error of 0.0016 using antithetic random variables, and an estimate of 1.4967 with an estimated relative error of 0.0022 using CMC. Hence, we can empirically observe the improvement of the antithetic random variable estimator over the CMC estimator. We used the following code.

```
clear all,format short g
seed=65;
rand('state',seed)
N=100000;
d=5;
u=[1,1,0.5,2,1.5]; % the nominal parameter vector

X_CMC=-log(rand(N,d)).*repmat(u,N,1); %sample for CMC
Score_CMC=H(X_CMC);
ell_CMC=mean(Score_CMC);
var_CMC=var(Score_CMC)/N;

rand('state',seed)
URVs=rand(N/2,d);  % get uniform random variables
X_ARV=-log([URVs;1-URVs]).*repmat(u,N,1); % sample for ARV
Score_ARV=H(X_ARV);
ell_ARV=mean(Score_ARV);
Covariance=cov(Score_ARV(1:N/2),Score_ARV(N/2+1:N));
var_ARV=(var(Score_ARV(1:N/2))+var(Score_ARV(N/2+1:N))+...
                    2*Covariance(1,2))/(2*N);

[ell_ARV,sqrt(var_ARV)/ell_ARV;ell_CMC,sqrt(var_CMC)/ell_CMC ]
```

The performance function H is defined in the following m-file:

```
% H.m
function shortest_path=H(X)
Path_1=X(:,1)+X(:,4);
Path_2=X(:,1)+X(:,3)+X(:,5);
Path_3=X(:,2)+X(:,3)+X(:,4);
Path_4=X(:,2)+X(:,5);
shortest_path=min([Path_1,Path_2,Path_3,Path_4],[],2);
```

5.3   Typical simulation output using CMC gives an estimate of 1.12 with an estimated relative error of 0.046, and with antithetic random variables gives an estimate of 1.13 with an estimated relative error of 0.034.

We used the following code.

```
clear all
seed= 12345; % use same seed for CMC and ARV
% CMC Method
rand('state',seed);
K=100;
M=10000;
N=99;
T=(M-K)/N;
R=rand(M,2);
A=-2*log(R(:,1)); % Interarrival time is Exp(1/2) distributed
S=1.5.*R(:,2)+0.5; % Service time is U[0.5,2] distributed
U=S(1:M-1)-A(2:M);
W=zeros(M,1);
for k=1:M-1
  W(k+1)=max(W(k)+U(k),0);
end
Y=zeros(N,1);
for k=1:N  % make the batches
  Y(k)=mean(W(K+1+T*(k-1):K+T*k));
end
ell=mean(Y);
varest=var(Y)/N; % variance of batch means estimator
RE = sqrt(varest)/ell; % RE for CMC method

% ARV Method
rand('state',seed);
Ma=5000;
Na=49;
R=rand(Ma,2);
R=[R;1-R]; % antithetic variables
A=-2*log(R(:,1)); % Interarrival time is Exp(1/2) distributed
S=1.5.*R(:,2)+0.5; % Service time is U[0.5,2] distributed
U=S(1:2*Ma-1)-A(2:2*Ma);
W1=zeros(Ma,1);
W2=zeros(Ma,1);
for k=1:Ma-1 % repeat the experiment twice with half the amount
            % of samples
  W1(k+1)=max(W1(k)+U(k),0);
  W2(k+1)=max(W2(k)+U(Ma+k),0);
end
for k=1:Na
  Y1(k)=mean(W1(K+1+T*(k-1):K+T*k));
  Y2(k)=mean(W2(K+1+T*(k-1):K+T*k));
end
Y=(Y1 + Y2)/2;
ella=mean(Y);
cova=cov(Y1,Y2);
varesta=(cova(1,1) + cova(2,2) + 2*cova(1,2))/(4*Na);
REa = sqrt(varesta)/ella; % RE for ARV method
[ell,RE; ella, REa] % output results
```

5.4 For this problem the control variables are

$$C_1 = X_1 + X_4$$
$$C_2 = X_1 + X_3 + X_5$$
$$C_3 = X_2 + X_3 + X_4$$
$$C_4 = X_2 + X_5 ,$$

which can be written as

$$C = \underbrace{\begin{pmatrix} 1 & 0 & 0 & 1 & 0 \\ 1 & 0 & 1 & 0 & 1 \\ 0 & 1 & 1 & 1 & 0 \\ 0 & 1 & 0 & 0 & 1 \end{pmatrix}}_{B} X .$$

The expectation of C is

$$r = \mathbb{E}[C] = \begin{pmatrix} 2 \\ 3 \\ 3 \\ 2 \end{pmatrix} .$$

Below, we estimate both Σ_C and σ_{XC} from the sample. However, note that Σ_C can be found explicitly and is given by

$$\Sigma_C = BB^T = \begin{pmatrix} 2 & 1 & 1 & 0 \\ 1 & 3 & 1 & 1 \\ 1 & 1 & 3 & 1 \\ 0 & 1 & 1 & 2 \end{pmatrix} .$$

Typical output from the program below gives an estimate of 1.18 with an estimated relative error of 0.020 for CMC, and an estimated relative error of 0.014 using the control variables.

```
N=1000;
u=[1,1,1,1,1]; % the nominal parameter

X=-log(rand(N,5)).*repmat(u,N,1); %samples for the CMC estimator
HX=H(X);
ell_CMC=mean(HX);
RE_CMC=std(HX)/ell_CMC/sqrt(N);

C1=X(:,1)+X(:,4);
C2=X(:,1)+X(:,3)+X(:,5);
C3=X(:,2)+X(:,3)+X(:,4);
C4=X(:,2)+X(:,5);
C=[C1,C2,C3,C4];

Sigma_C=cov(C);

temp=cov(HX,C1);
sigma_XC(1,1)=temp(1,2);
```

```
temp=cov(HX,C2);
sigma_XC(2,1)=temp(1,2);
temp=cov(HX,C3);
sigma_XC(3,1)=temp(1,2);
temp=cov(HX,C4);
sigma_XC(4,1)=temp(1,2);

alpha = inv(Sigma_C)*sigma_XC;

R=repmat([2,3,3,2],N,1);
HXstar=HX - (alpha'*(C'-R'))';

ell_Control=mean(HXstar);
var_Control = var(HX) - sigma_XC'*inv(Sigma_C)*sigma_XC ;
RE_Control = sqrt(var_Control)/sqrt(N)/ell_Control;
[ell_CMC,RE_CMC;ell_Control,RE_Control]
```

5.5   We use C_4 in **Example 5.5** as a control variable for the estimation $\mathbb{E}[W_4]$. The expectation of C_4 is $3 \times (1.25 - 2) = 2.25$. Figure 15.1 illustrates the strong correlation between the random variables W_4 and C_4. The sample correlation here is $\varrho = 0.6$, based on a sample of size 1,000.

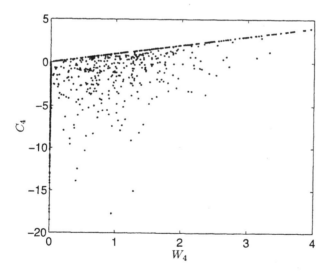

Figure 15.1   The random variables W_4 and C_4 are positively correlated.

Typical simulation output gives a point estimate for $\mathbb{E}[W_4]$ of 0.69 with an estimated relative error of 0.03, giving a 95% confidence interval of $(0.65, 0.74)$. The estimated relative error is around 0.04 using CMC. We used the following code.

```
N=1000;
A=-2*log(rand(N,4)); % Interarrival time is Exp(1/2) distributed
S=1.5.*rand(N,4)+0.5; % Service time is U[0.5,2] distributed
U=S(:,1:3)-A(:,2:4);
W=zeros(N,4);
C=zeros(N,4);
for k=1:3
  W(:,k+1)=max(W(:,k)+U(:,k),0);
  C(:,k+1)=C(:,k)+U(:,k);
end
RE_CMC = sqrt(var(W(:,4)))/sqrt(N)/mean(W(:,4))

cov_XC=cov(W(:,4),C(:,4));
var_C=var(C(:,4));

alpha=cov_XC(1,2)/var_C;
r = -2.25; %expected value of C_4
Y=W(:,4)-alpha.*(C(:,4)-r);

ell_Control=mean(Y);
var_Control=var(W(:,4))-2*alpha*cov_XC(1,2)+alpha*alpha*var_C;
RE_Control = sqrt(var_Control)/sqrt(N)/ell_Control
ell_Control
[ell_Control-1.96*sqrt(var_Control/N),ell_Control+...
                          1.96*sqrt(var_Control/N)]
corr(W(:,4),C(:,4))
plot(W(:,4),C(:,4),'.')
```

5.6   Using the hint, we write

$$\text{Var}(U) = \mathbb{E}[U^2] - (\mathbb{E}[U])^2 = \mathbb{E}[\,\mathbb{E}[U^2\,|\,V]\,] - (\mathbb{E}[\,\mathbb{E}[U\,|\,V]\,])^2$$
$$= \mathbb{E}[\,\mathbb{E}[U^2\,|\,V]\,] - (\mathbb{E}[\,\mathbb{E}[U\,|\,V]\,])^2 + \mathbb{E}[\mathbb{E}[U\,|\,V]^2] - \mathbb{E}[\mathbb{E}[U\,|\,V]^2]$$
$$= \underbrace{\mathbb{E}[\,\mathbb{E}[U^2\,|\,V]\,] - \mathbb{E}[\mathbb{E}[U\,|\,V]^2]}_{\mathbb{E}[\,\text{Var}(U\,|\,V)\,]} + \underbrace{\mathbb{E}[\mathbb{E}[U\,|\,V]^2] - (\mathbb{E}[\,\mathbb{E}[U\,|\,V]\,])^2}_{\text{Var}(\,\mathbb{E}[U\,|\,V]\,)} .$$

Hence,

$$\text{Var}(U) = \mathbb{E}[\,\text{Var}(U\,|\,V)\,] + \text{Var}(\,\mathbb{E}[U\,|\,V]\,) .$$

5.7 (a) Let

$$M_X(t) = \mathbb{E}[e^{tX}] = \int_0^\infty e^{tx} \lambda e^{-\lambda x}\, dx = \lambda/(\lambda - t), \quad t < -\lambda$$

be the moment generating function of $X \sim \text{Exp}(\lambda)$ and let

$$G(z) = \mathbb{E}[z^R] = \sum_{r=1}^\infty z^r (1-p)^{r-1} p = \frac{p}{1-p} \sum_{r=1}^\infty z^r (z(1-p))^r = \frac{zp}{1 - (1-p)z}, \quad |z| \leqslant 1$$

be the probability generating function of $R \sim G(p)$. Then, the moment generation function, say $M(t)$, of the random sum S_R is given by

$$M(t) = \mathbb{E}\left[e^{tS_R}\right] = \mathbb{E}\left[e^{t\sum_{i=1}^{R} X_i}\right]$$

$$= \mathbb{E}\left[\mathbb{E}\left[e^{t\sum_{i=1}^{R} X_i} \mid R\right]\right] = \mathbb{E}\left[(\mathbb{E}[e^{tX}])^R\right]$$

$$= \mathbb{E}\left[M_X(t)^R\right] = G(M_X(t))$$

$$= \frac{\frac{\lambda}{\lambda-t}p}{1-(1-p)\frac{\lambda}{\lambda-t}}$$

$$= \frac{\lambda p}{\lambda - t - (1-p)\lambda} = \frac{\lambda p}{\lambda p - t},$$

which is the moment generating function of the $\text{Exp}(\lambda p)$ distribution. Hence, $S_R \sim \text{Exp}(\lambda p)$.

(b) Typical simulation output gives an estimate for $\mathbb{P}(S_R > 10)$ of 0.37 using CMC, with an estimated relative error of 0.042.

(c) The conditional estimator here is $\widehat{\ell}_c = \frac{1}{N}\sum_{k=1}^{N} F_k$, where

$$F_k = \begin{cases} \exp\left(-10 + \sum_{i=2}^{R_k} X_{ki}\right) & \text{if } \sum_{i=2}^{R_k} X_{ki} < 10 \\ 1 & \text{otherwise.} \end{cases}$$

Typical output gives an estimate for $\mathbb{P}(S_R > 10)$ of 0.37 using conditional Monte Carlo, with an estimated relative error of 0.040. Part (a) shows that the true probability is $e^{-1} \approx 0.36788$. In this case the conditional estimator gives only a very small improvement upon the CMC estimator. For larger sample sizes, for example $N = 100,000$, we consistently obtained an improvement in relative error of only 2–3%.

5.8   Let $p_i = \mathbb{P}(R = i) = (1-p)^{i-1}p$ for $i = 1, 2, \ldots, 7$ and $p_8 = 1 - \sum_{k=1}^{7} p_k$. With $N_i = p_i N$, the sample sizes are 2,500, 1,875, 1,406, 1,055, 791, 593, 445, and 1,335. Typical simulation output gives an estimate of 0.082 with an estimated relative error of 0.025. The true probability is $e^{-2.5} \approx 0.0821$.

For the case $N_i \propto N p_i \sigma_i$, we estimate the σ_i via a pilot run of size 1,000. Typical strata sizes $\{N_i\}$ are now 0, 0, 0, 936, 809, 985, 1,177, and 6,093. The relative error is significantly lower than before — around 0.015. A typical 95% confidence interval is $(0.079, 0.084)$. The following script was used for the second method.

```
clear all
N=10000;
Nsig=1000;
p=0.25;
lambda=1;
n=8;

bp=p.*[1,(1-p),(1-p)^2,(1-p)^3,(1-p)^4,(1-p)^5,(1-p)^6,inf];
bp(n)=1-sum(bp(1:n-1));
% N_i \propto N \sigma_i p_i
% \sigmas are estimated using a sample size of Nsig
```

```
Ns=round(Nsig.*bp);

for k=1:(n-1)
 X=-log(rand(Ns(k),k))./lambda;
 S=sum(X,2);
 HC(k)=mean(S>10);
 V(k)=var(S>10);
end

S=[];
for j=1:Ns(n)
 R=geornd(p)+8;
 X=-log(rand(R,1))./lambda;
 S(j)=sum(X);
end
HC(n)=mean(S>10);
V(n)=var(S>10);

% V now contains our pilot estimates for \sigma_i^2


No=N.*bp.*sqrt(V)./sum(bp.*sqrt(V)); Ns=round(No);
diff=N-sum(Ns);

if diff>0
 Nsort=No-Ns;
 [Nsort,idx]=sort(Nsort,'descend');
 for k=1:diff
  Ns(idx(k))=Ns(idx(k))+1;
 end
end

if sum(Ns)~=N
 error('The sum of the strata N_i is not equal to N!')
end

for k=1:(n-1)
 X=-log(rand(Ns(k),k))./lambda;
 S=sum(X,2);
 if Ns(k)>0
  HC(k)=mean(S>10);
  V(k)=var(S>10);
 else
HC(k)=0;
V(k)=0;
 end
end

S=[];
for j=1:Ns(n)
 R=geornd(p)+8;
 X=-log(rand(R,1))./lambda;
 S(j)=sum(X);
```

```
end
HC(n)=mean(S>10);
V(n)=var(S>10);

ell_opt=sum(bp.*HC)
Nidx=Ns>0;
var_ell_opt=sum(bp(Nidx).^2.*V(Nidx)./Ns(Nidx));
RE = sqrt(var_ell_opt)/ell_opt
[ell_opt*(1 - 1.96*RE), ell_opt*(1 + 1.96*RE)]
```

5.9   The optimization problem is

$$\min_{N_1\ldots,N_m} \sum_{i=1}^m \frac{a_i^2}{N_i} \ \ \text{subject to} \ \ N_1 + \cdots + N_m = N \ ,$$

where $a_i = p_i \sigma_i$ for $i = 1, 2, \ldots, m$. To solve it, construct the Lagrangian

$$\mathcal{L}(N_1, \ldots, N_m, \lambda) = \sum_{i=1}^m \frac{a_i^2}{N_i} + \lambda \sum_{i=1}^m N_i \ .$$

The solution to $\frac{\partial}{\partial N_i}\mathcal{L} = 0$ is $N_i^* = \frac{|a_i|}{\sqrt{\lambda}}$ for each i. Hence, summing over all i, we obtain $\sqrt{\lambda^*} = \frac{1}{N} \sum_{i=1}^m |a_i|$. Therefore, $N_i^* = \frac{|a_i|}{\sum_{i=1}^m |a_i|} N$, which justifies the stratified sampling scheme.

5.10   Using $\mathbf{p} = (0.3, 0.1, 0.8, 0.1, 0.2)$ with the permutation Monte Carlo **Algorithm 5.4.2**, we obtained an estimated *un*reliability of 0.906 with an estimated relative error of 0.0028. The true unreliability is 0.907240. For $\mathbf{p} = (0.95, 0.95, 0.95, 0.95, 0.95)$ the estimated unreliability was found to be $5.25 \cdot 10^{-3}$ with a relative error of 0.037. The true unreliability here is $5.21938 \cdot 10^{-3}$.

5.11   Using $\mathbf{p} = (0.3, 0.1, 0.8, 0.1, 0.2)$ with the min-cut **Algorithm 5.4.3** we obtained an estimated *un*reliability of 0.923 with an estimated relative error of 0.01. For $\mathbf{p} = (0.95, 0.95, 0.95, 0.95, 0.95)$ the estimated unreliability was found to be $5.22 \cdot 10^{-3}$ with an estimated relative error of 0.001. We observe that the min-cut method works better than the permutation Monte Carlo method when the unreliabilities are small. If this is not the case, the permutation Monte Carlo method is better. The following code was used for the min-cut method.

```
function [unrel,RE,time] = min_cut(p,N)

q = 1 - p;
n = length(p); % Number of links.
C = [1 2 0; 1 3 5; 2 3 4; 4 5 0];
a = [q(1)*q(2) q(1)*q(3)*q(5) q(2)*q(3)*q(4) q(4)*q(5)];
a_sum = sum(a); A = a./a_sum;

tic;
for k = 1:N
    % Generate a discrete random variable J.
    j = rand;
```

```
    if j <= A(1)
        J = 1;
    elseif j > A(1) & j <= sum(A(1:2))
        J = 2;
    elseif j > sum(A(1:2)) & j <= sum(A(1:3))
        J = 3;
    else
        J = 4;
    end

    % Generate a vector X.
    index1 = setdiff(C(J,:),0);
    index2 = setxor(index1,1:5);
    X(1,index1) = 0;
    X(1,index2) = (rand(1,length(index2)) <= p(index2));

    % Compute an estimated unreliability.
    S = 0;
    for i = 1:4
        index = setdiff(C(i,:),0);
        S = S + prod(1-X(index));
    end
    prob(k) = a_sum/S;
end

unrel = mean(prob);
sd = sqrt((sum((prob - unrel).^2))/N);
RE = sd/sqrt(N)/unrel; % Relative error.

time = toc;
```

5.12 (a) We need to show that

$$H(\mathbf{x}) = \max_k \prod_{i \in A_k} x_i = 1 - \prod_{k=1}^{m} \left(1 - \prod_{i \in A_k} x_i\right). \tag{15.1}$$

First, from the description of A_k as the k-th minimal path set, we note that the product $\prod_{i \in A_k} x_i$ is equal to 1 if and only if every element in A_k is functioning, that is, one of the components in at least one minimal path is working, and is 0 otherwise. Second, the maximum of these products over all minimal path sets is 1 if and only if the system is functioning, and is 0 otherwise. This proves the first equality in (15.1). The second equality follows from the fact that

$$1 - \prod_{k=1}^{m} \left(1 - \prod_{i \in A_k} x_i\right)$$

is 1 if at least one of the products $\prod_{i \in A_k} x_i$ is equal to 1 (that is, the system is functioning), and is 0 otherwise.

(b) Define

$$Y_k = \prod_{i \in A_k} X_i, \quad k = 1, \ldots, m$$

as the indicator of the event that all components in A_k are functioning and let $S = \sum_{k=1}^{m} Y_k$.

We wish to estimate the reliability of the system, that is, $r = \mathbb{P}(S > 0)$, using **Proposition 5.4**. Define

$$b_k = \mathbb{P}(Y_k = 1) = \mathbb{P}\left(\prod_{i \in A_k} X_i = 1\right) = \prod_{i \in A_k} \mathbb{P}(X_i = 1) = \prod_{i \in A_k} p_i ,$$

and let $b = \sum_{j=1}^m b_j$. By analogy to **Algorithm 5.4.3** we obtain the following algorithm.

Algorithm (Conditioning via Paths).

1. *Generate a discrete random variable K with $\mathbb{P}(K = k) = b_k/b$, $k = 1, \ldots, m$.*

2. *Set X_i equal to 1 for all $i \in A_K$ and generate the values of all other X_i, $i \notin A_K$ from their corresponding $\mathsf{Ber}(p_i)$ distributions.*

3. *Evaluate b/S, where S denotes the number of minimal paths that have all their components functioning.*

4. *Repeat Steps 1–3 N times and take $N^{-1} \sum_{i=1}^N b/S_i$ as an estimator of $r = \mathbb{P}(S > 0)$.*

(c) Using the same **p** as in Problems 5.10 and 5.11, we obtained the following results for the path conditioning algorithm using a sample size of $N = 1,000$. For $\mathbf{p} = (0.3, 0.1, 0.8, 0.1, 0.2)$ we obtained an estimated reliability of 0.0925 (that is, an unreliability of 0.9075) with an estimated relative error of 0.0082. For $\mathbf{p} = (0.95, 0.95, 0.95, 0.95, 0.95)$ the estimated reliability was found to be 0.9934 with an estimated relative error of 0.01. Clearly the min-path algorithm is better suited toward estimating small system reliabilities. For example, using $\mathbf{p} = (0.05, 0.05, 0.05, 0.05, 0.05)$ the system reliability is estimated as 0.0052 with an estimated relative error of 0.002. The true reliability is $5.219375 \cdot 10^{-3}$ in this case.

5.13   First, suppose that $H(\mathbf{x}) \geqslant 0$. In this case $|H(\mathbf{x})| = H(\mathbf{x})$ and

$$\int |H(\mathbf{x})| \, f(\mathbf{x}) \, d\mathbf{x} = \int H(\mathbf{x}) \, f(\mathbf{x}) \, d\mathbf{x} = \ell .$$

The purported minimum-variance importance sampling density here is

$$g^*(\mathbf{x}) = \frac{|H(\mathbf{x})| \, f(\mathbf{x})}{\int |H(\mathbf{x})| \, f(\mathbf{x}) \, d\mathbf{x}} = \frac{H(\mathbf{x}) \, f(\mathbf{x})}{\ell} ,$$

yielding

$$\mathrm{Var}_{g^*}\left(H(\mathbf{X}) \frac{f(\mathbf{X})}{g^*(\mathbf{X})}\right) = \mathrm{Var}_{g^*}\left(H(\mathbf{X}) \frac{f(\mathbf{X})}{H(\mathbf{X}) \, f(\mathbf{X})/\ell}\right) = 0 .$$

Hence, when $H(\mathbf{x}) \geqslant 0$, g^* does indeed give minimal variance, since the variance is always nonnegative.

Now suppose that $H(\mathbf{x})$ is arbitrary. We know that

$$\min_g \mathrm{Var}_g\left(H(\mathbf{X}) \frac{f(\mathbf{X})}{g(\mathbf{X})}\right)$$

is equivalent to

$$\min_g \mathbb{E}_g\left[\left(H(\mathbf{X}) \frac{f(\mathbf{X})}{g(\mathbf{X})}\right)^2\right],$$

which in turn is equivalent to

$$\min_{g} \mathbb{E}_g \left[\left(|H(\mathbf{X})| \frac{f(\mathbf{X})}{g(\mathbf{X})} \right)^2 \right].$$

From the first part of the problem we know that the solution to this minimization problem is

$$g^*(\mathbf{x}) = \frac{|H(\mathbf{x})| f(\mathbf{x})}{\int |H(\mathbf{x})| f(\mathbf{x}) \, d\mathbf{x}},$$

which had to be shown.

5.14   Let $Z \sim N(0, 1)$. We wish to estimate $\mathbb{P}(Z > 4)$ via importance sampling using the shifted exponential sampling pdf

$$g(x) = e^{-(x-4)}, \quad x \geq 4,$$

and choosing N large enough to obtain at least three significant digits of accuracy.
    The true value is

$$\mathbb{P}(Z > 4) = \int_4^\infty \frac{e^{-\frac{1}{2}x^2}}{\sqrt{2\pi}} \, dx = \frac{1}{2} - \frac{1}{2} \operatorname{erf}(2\sqrt{2}) \approx 0.00003167124.$$

The code below, with sample size $N = 2 \cdot 10^6$, gave an estimate of $3.165 \cdot 10^{-5}$ with an estimated relative error of 0.00085.

```
N=2000000;
X=4-log(rand(N,1));
w=exp(-0.5.*(X.^2)).*exp(X-4)./sqrt(2*pi);
ell=mean(w)
re=std(w)/(sqrt(N)*ell)
```

5.15   The *Pearson χ^2 discrepancy measure* (see **Remark 1.14.1**) is given by

$$d(g, h) = \frac{1}{2} \int \frac{[g(\mathbf{x}) - h(\mathbf{x})]^2}{h(\mathbf{x})} \, d\mathbf{x}.$$

Consider the minimization problem

$$\min_{g} d(g^*, g), \tag{15.2}$$

where g^* is the zero-variance importance sampling pdf $g^*(\mathbf{x}) = H(\mathbf{x}) f(\mathbf{x})/\ell$, with $\ell = \mathbb{E}_f[H(\mathbf{X})]$ and assuming $H(\mathbf{x}) \geq 0$. Now, we have

$$\begin{aligned}
d(g^*, g) &= \frac{1}{2} \int \frac{[g^*(\mathbf{x}) - g(\mathbf{x})]^2}{g(\mathbf{x})} \, d\mathbf{x} \\
&= \frac{1}{2} \int \frac{[g^*(\mathbf{x})^2 - 2g^*(\mathbf{x})g(\mathbf{x}) + g(\mathbf{x})^2]}{g(\mathbf{x})} \, d\mathbf{x} \\
&= \frac{1}{2} \int \left[\frac{g^*(\mathbf{x})^2}{g(\mathbf{x})} - 2g^*(\mathbf{x}) + g(\mathbf{x}) \right] d\mathbf{x} \\
&= -\frac{1}{2} + \frac{1}{2} \int \left[\frac{g^*(\mathbf{x})^2}{g(\mathbf{x})} \right] d\mathbf{x}.
\end{aligned}$$

Hence, (15.2) is equivalent to the minimization problem

$$\min_{g} \mathbb{E}_g \left[\left(\frac{g^*(\mathbf{X})}{g(\mathbf{X})} \right)^2 \right] ,$$

which in turn is equivalent to

$$\min_{g} \mathbb{E}_g \left[\left(\frac{H(\mathbf{X}) f(\mathbf{X})}{g(\mathbf{X})} \right)^2 \right] ,$$

which is equivalent to the variance minimization problem **(5.44)**, that is,

$$\min_{g} \operatorname{Var}_g \left(H(\mathbf{X}) \frac{f(\mathbf{X})}{g(\mathbf{X})} \right) , \tag{15.3}$$

as was to be shown. Note that the problems are still equivalent when $H(\mathbf{x}) \not\geq 0$ and g^* is the corresponding minimum-variance importance sampling density.

5.16   Since the components of the random vector $\mathbf{X} = (X_1, \ldots, X_5)$ are independent and are distributed according to a one-parameter exponential family parameterized by the mean, the CE updating formulas are given by **(5.70)**, that is,

$$\widehat{v}_i = \frac{\sum_{k=1}^{N} H(\mathbf{X}_k) X_{ki}}{\sum_{k=1}^{N} H(\mathbf{X}_k)}, \quad i = 1, \ldots, 5 . \tag{15.4}$$

To obtain a good reference vector we evaluate (15.4) from a pilot sample of size 1,000. A typical result is

$$\widehat{\mathbf{v}} = (1.177, 1.212, 0.505, 2.628, 2.010) .$$

Performing the actual estimation of ℓ with a sample size of 10^5, we found an estimate $\widehat{\ell} = 1.4992$ with an estimated relative error of 0.0014, which is significantly smaller than the relative errors using CMC or antithetic random variables in Problem 5.2. The following m-file was used.

```
%main.m
clear all
u=[1,1,0.5,2,1.5]; % the nominal parameter
N = 1000;
% Estimate CE optimal v
Y =-log(rand(N,5)).*repmat(u,N,1);
HY = repmat(H(Y),1,5).*Y;
v = mean(HY)/mean(H(Y))
N = 100000;
% Actual estimation
X = -log(rand(N,5)).*repmat(v,N,1);
W = f(X,u)./f(X,v);
HW = H(X).*W;
ell = mean(HW)
RE = std(HW)/sqrt(N)/ell
```

The functions H.m and f.m are defined as:

```
function shortest_path=H(X)
Path_1=X(:,1)+X(:,4);
Path_2=X(:,1)+X(:,3)+X(:,5);
Path_3=X(:,2)+X(:,3)+X(:,4);
Path_4=X(:,2)+X(:,5);
shortest_path=min([Path_1,Path_2,Path_3,Path_4],[],2);
```

and

```
function y = f(x,v)
N= size(x,1);
V = repmat(v,N,1);
y = prod(exp(-x./V)./V,2);
```

5.17   A pdf belonging to the natural exponential family can be written (see **(A.9)**) as

$$f(\mathbf{x}; \boldsymbol{\theta}) = c(\boldsymbol{\theta}) \, e^{\boldsymbol{\theta} \cdot \mathbf{t}(\mathbf{x})} \, h(\mathbf{x}) \,, \tag{15.5}$$

where $\mathbf{t}(\mathbf{x}) = (t_1(\mathbf{x}), \ldots, t_m(\mathbf{x}))^T$, $\boldsymbol{\theta} = (\theta_1, \ldots, \theta_m)^T$, $t_i(\mathbf{x})$, $h(\mathbf{x})$ and $c(\boldsymbol{\theta}) > 0$, and $\boldsymbol{\theta} \cdot \mathbf{t}(\mathbf{x})$ is the inner product $\sum_{i=1}^m \theta_i t_i(\mathbf{x})$. With a pdf of this form, equation **(5.62)** becomes

$$\mathbb{E}_{\boldsymbol{\theta}_0} \left[H(\mathbf{X}) \, \nabla \ln \left(c(\boldsymbol{\theta}) \, e^{\boldsymbol{\theta} \cdot \mathbf{t}(\mathbf{X})} \, h(\mathbf{X}) \right) \right] = 0 \,,$$

which can be expanded to

$$\mathbb{E}_{\boldsymbol{\theta}_0} \left[H(\mathbf{X}) \, \nabla \left(\ln \left(c(\boldsymbol{\theta}) \right) + \boldsymbol{\theta} \cdot \mathbf{t}(\mathbf{X}) + \ln \left(h(\mathbf{X}) \right) \right) \right] = 0 \,.$$

Applying the gradient operator, ∇, yields

$$\mathbb{E}_{\boldsymbol{\theta}_0} \left[H(\mathbf{X}) \left(\frac{\frac{\partial}{\partial \theta_k} c(\boldsymbol{\theta})}{c(\boldsymbol{\theta})} + t_k(\mathbf{X}) \right) \right] = 0 \quad k = 1, \ldots, m \,,$$

which can be written compactly as

$$\mathbb{E}_{\boldsymbol{\theta}_0} \left[H(\mathbf{X}) \left(\frac{\nabla c(\boldsymbol{\theta})}{c(\boldsymbol{\theta})} + \mathbf{t}(\mathbf{X}) \right) \right] = 0 \,. \tag{15.6}$$

5.18   Applying (15.6), we have that the CE optimal λ must satisfy

$$\mathbb{E}_{\lambda_0} \left[H(X) \left(-\frac{1}{\lambda} + X \right) \right] = 0 \,.$$

Hence,

$$\frac{\mathbb{E}_{\lambda_0}[H(X)]}{\lambda} = \mathbb{E}_{\lambda_0}[H(X) \, X] \,,$$

leading to

$$\lambda^* = \frac{\mathbb{E}_{\lambda_0}[H(X)]}{\mathbb{E}_{\lambda_0}[H(X) \, X]} \,. \tag{15.7}$$

Alternatively, we can derive the optimal parameter from (**A.15**), that is, from

$$v^* = \frac{\mathbb{E}_u[H(X)\,X]}{\mathbb{E}_u\,[H(X)]}\,. \tag{15.8}$$

Namely, if we reparameterize the $\mathsf{Exp}(\lambda)$ distribution via $\lambda = 1/v$, where $v = 1/\lambda$ corresponds to the mean, then the optimal v is given by (15.8). Therefore, the optimal λ is given by $\lambda = 1/v^*$, which is in accordance with (15.7). (Note that $\mathbb{E}_u$ means that the expectation is taken with respect to the nominal parameter $u = 1/\lambda_0$.)

We can estimate λ^*, for example, by drawing a random sample $X_1, \ldots, X_N$ from $\mathsf{Exp}(\lambda_0)$ and evaluating

$$\widehat{\lambda^*} = \frac{\sum_{k=1}^{N} H(X_k)}{\sum_{k=1}^{N} H(X_k)\,X_k}\,.$$

5.19   Using **Table A.1**, we can represent the $\mathsf{Weib}(\alpha, \lambda)$ distribution (with fixed α) as a one-parameter exponential family of the form (15.5), with $t(x) = x^\alpha$ and $c(\theta) = -\theta\,\alpha$, where $\theta = -\lambda^\alpha$. Hence, $\nabla c(\theta) = -\alpha$. Thus, (15.6) yields

$$\mathbb{E}_{\theta_0}\left[H(X)\left(\frac{1}{\theta_1} + X^\alpha\right)\right] = 0\,.$$

Substituting back λ_0 and λ, the last equation becomes

$$\mathbb{E}_{\lambda_0}\left[H(X)\left(-\lambda^{-\alpha} + X^\alpha\right)\right] = 0\,.$$

Rearranging, we obtain

$$\lambda^{-\alpha}\mathbb{E}_{\lambda_0}\left[H(X)\right] = \mathbb{E}_{\lambda_0}\left[H(X)X^\alpha\right],$$

which becomes

$$\lambda = \left(\frac{\mathbb{E}_{\lambda_0}\left[H(X)\right]}{\mathbb{E}_{\lambda_0}\left[H(X)X^\alpha\right]}\right)^{1/\alpha}\,.$$

5.20   The CE optimal parameter θ^* is the solution to the equation

$$\mathbb{E}\left[I_{\{S(\mathbf{X})\geqslant\gamma\}}\,\nabla\ln f(\mathbf{X};\theta)\right] = 0\,,$$

with

$$f(\mathbf{x};\theta) = \prod_{k=1}^{n}\theta\,e^{-\theta x_k}.$$

Since

$$\ln f(\mathbf{x};\theta) = \sum_{k=1}^{n}\ln\left(\theta\,e^{-\theta x_k}\right) = \sum_{k=1}^{n}(\ln\theta - \theta x_k) = n\ln\theta - \theta\sum_{k=1}^{n}x_k\,,$$

the equation to find the CE optimal parameters is given by

$$\mathbb{E}\left[I_{\{S(\mathbf{X})\geqslant\gamma\}}\left(\frac{n}{\theta} - \sum_{k=1}^{n}X_k\right)\right] = 0\,.$$

Rearranging, we have

$$\frac{n}{\theta} \, \mathbb{E}\left[I_{\{S(\mathbf{X}) \geqslant \gamma\}}\right] = \mathbb{E}\left[I_{\{S(\mathbf{X}) \geqslant \gamma\}} \sum_{k=1}^{n} X_k\right] \, ,$$

which becomes

$$\theta = \frac{\mathbb{E}\left[I_{\{S(\mathbf{X}) \geqslant \gamma\}}\right]}{\mathbb{E}\left[I_{\{S(\mathbf{X}) \geqslant \gamma\}} \, \overline{X}\right]} \, .$$

Thus, the result is shown.

5.21 (a) Let $Z \sim \text{Exp}(1)$ and $G(z) = z^{1/\alpha}/\lambda_0$. Then, for any $z \geqslant 0$,

$$\mathbb{P}(G(Z) \leqslant z) = \mathbb{P}(Z^{1/\alpha}/\lambda_0 \leqslant z) = \mathbb{P}\left(Z \leqslant (\lambda_0 z)^\alpha\right) = 1 - e^{-(\lambda_0 z)^\alpha} \, ,$$

which corresponds to the cdf of the Weib(α, λ_0) distribution.

(b) Let $X \sim \text{Weib}(\alpha, \lambda)$. We wish to estimate $\ell = \mathbb{E}[H(X)]$ via the TLR method. Recalling that $\lambda^{-1} Z^{1/\alpha}$ is Weib(α, λ) distributed if $Z \sim \text{Exp}(1)$, we have that $X = G(Z)$ and so $H(X) = H(G(Z)) = \tilde{H}(Z)$, where $\tilde{H}(Z) = H(\lambda^{-1} Z^{1/\alpha})$. Thus, $\ell = \mathbb{E}[\tilde{H}(Z)] = \mathbb{E}[H(\lambda^{-1} Z^{1/\alpha})]$. We could take $h(z; \theta) = \theta \, e^{-\theta z}$ as the sampling pdf, in which case the TLR estimator becomes

$$\hat{\ell} = \frac{1}{N} \sum_{k=1}^{N} \lambda^{-1} \, Z_k^{1/\alpha} \, \widetilde{W}(Z_k; 1, \theta) \, , \tag{15.9}$$

with

$$\widetilde{W}(Z_k; 1, \theta) = \frac{h(Z_k; 1)}{h(Z_k; \theta)} = \frac{e^{-Z_k}}{\theta \, e^{-\theta Z_k}}$$

and $Z_k \sim \text{Exp}(\theta)$.

(c) To find the optimal parameter vector θ^* of the TLR estimator we can solve the following CE program:

$$\max_{\theta} D(\theta) = \max_{\theta} \mathbb{E}_\eta \left[\tilde{H}(Z) \, \widetilde{W}(Z; 1, \eta) \ln h(Z; \theta)\right] \, .$$

Writing this out, we wish to maximize

$$\mathbb{E}_\eta \left[\tilde{H}(Z) \, \frac{e^{-Z}}{\eta \, e^{-\eta Z}} \ln\left(\theta \, e^{-\theta Z}\right)\right]$$

over θ. Assuming that we can change the order of differentiation and expectation, this is equivalent to solving

$$\mathbb{E}_\eta \left[\tilde{H}(Z) \, e^{-(1-\eta)Z} \left(\frac{1}{\theta} - Z\right)\right] = 0 \, ,$$

which yields

$$\theta = \frac{\mathbb{E}_\eta \left[\tilde{H}(Z) \, e^{-(1-\eta)Z}\right]}{\mathbb{E}_\eta \left[Z \tilde{H}(Z) \, e^{-(1-\eta)Z}\right]} \, ,$$

or equivalently

$$\theta = \frac{\mathbb{E}_\eta \left[\tilde{H}(Z) \, \widetilde{W}(Z; 1, \eta)\right]}{\mathbb{E}_\eta \left[Z \tilde{H}(Z) \, \widetilde{W}(Z; 1, \eta)\right]} \, .$$

5.22   We are given $\ell = \sum_{i=1}^{m} a_i \ell_i$, where $\ell_i = \int H_i(\mathbf{x}) \, d\mathbf{x}$, the $\{a_i\}$ are known coefficients, and $Q(\mathbf{x}) = \sum_{i=1}^{m} a_i H_i(\mathbf{x})$. Thus,

$$\ell = \mathbb{E}_g \left[\frac{Q(\mathbf{X})}{g(\mathbf{X})} \right] = \int Q(\mathbf{x}) \, d\mathbf{x}$$

for any pdf g such that $Q(\mathbf{x}) = 0$ implies that $g(\mathbf{x}) = 0$ (g dominates Q). Thus, $L = L(\mathbf{X}) = Q(\mathbf{X})/g(\mathbf{X})$ is an unbiased estimator of ℓ. Moreover, writing $\ell = \mathbb{E}_g[L(\mathbf{X})]$, we know from **5.45** (see also Problem 5.13) that the minimum-variance importance sampling density is given by

$$g^*(\mathbf{x}) = \frac{|L(\mathbf{x})| \, g(\mathbf{x})}{\int |L(\mathbf{x})| \, g(\mathbf{x}) \, d\mathbf{x}} = \frac{|Q(\mathbf{x})|}{\int |Q(\mathbf{x})| \, d\mathbf{x}} .$$

This gives a minimum variance of $\mathbb{E}_{g^*}[L^2(\mathbf{X})] - \ell^2$, where

$$\mathbb{E}_{g^*}[L^2(\mathbf{X})] = \int \frac{Q^2(\mathbf{x})}{g^*(\mathbf{x})} \, d\mathbf{x} = \int \frac{Q^2(\mathbf{x})}{|Q(\mathbf{x})|/\int |Q(\mathbf{x})| \, d\mathbf{x}} \, d\mathbf{x} = \left(\int |Q(\mathbf{x})| \, d\mathbf{x} \right)^2 ,$$

as had to be shown.

5.23   It is important to realize that here $\ell = b \, \mathbb{E}[H(X)]$. The CMC estimator is given by

$$\widehat{\ell} = \frac{b}{N} \sum_{i=1}^{N} H(X_i) .$$

Its variance is

$$\text{Var}(\widehat{\ell}) = \frac{b^2}{N} \text{Var}\,(H(X)) = \frac{b^2}{N} \left(\mathbb{E}[H(X)^2] - \ell^2/b^2 \right) = \frac{b^2 \, \mathbb{E}[H(X)^2] - \ell^2}{N} .$$

Since $H(x) \leqslant c$,

$$\text{Var}(\widehat{\ell}) \leqslant \frac{b^2 \, \mathbb{E}[cH(X)] - \ell^2}{N} = \frac{bc\ell - \ell^2}{N} = \frac{\ell\,(bc - \ell)}{N} .$$

The hit-or-miss estimator is given by

$$\widehat{\ell}^h = \frac{bc}{N} \sum_{i=1}^{N} I_{\{Y_i < H(X_i)\}} .$$

Its variance is

$$\text{Var}(\widehat{\ell}^h) = \frac{b^2 c^2}{N} \text{Var}\,\left(I_{\{Y < H(X)\}} \right) = \frac{b^2 c^2}{N} \, \mathbb{P}(Y \leqslant H(X)) \, (1 - \mathbb{P}(Y \leqslant H(X)))$$
$$= \frac{b^2 c^2}{N} \frac{\ell}{bc} \left(1 - \frac{\ell}{bc} \right) = \frac{\ell\,(bc - \ell)}{N} .$$

Hence, the variance of the hit-or-miss estimator is no less than the variance of the CMC estimator.

CHAPTER 16

MARKOV CHAIN MONTE CARLO

6.1   We wish to establish the detailed balance equation **(6.3)**:

$$\pi_i \, p_{ij} = \pi_j \, p_{ji}, \quad i, j \in \mathscr{X}.$$

Denote $\beta_{ij} = \pi_i \, p_{ij}$ to be the left-hand side of the detailed balance equation. Then, the right-hand side is β_{ji} and we have to establish that $\beta_{ij} = \beta_{ji}$. We have two major cases to consider:

Case 1, where $i = j$. This case follows directly from the definition. We have $\beta_{ij} = \beta_{ii} = \pi_i \, p_{ii} = \pi_j \, p_{jj} = \beta_{jj} = \beta_{ji}$.

Case 2, where $i \neq j$. Here, $\beta_{ij} = \pi_i \, p_{ij} = \pi_i \, q_{ij} \, \alpha_{ij} = \pi_i \, q_{ij} \, \min\left\{\frac{\pi_j q_{ji}}{\pi_i q_{ij}}, 1\right\}$. Hence,

$$\beta_{ij} = \begin{cases} \pi_j \, q_{ji} & \text{if} \quad \frac{\pi_j \, q_{ji}}{\pi_i \, q_{ij}} \leqslant 1 \\ \pi_i \, q_{ij} & \text{if} \quad \frac{\pi_j \, q_{ji}}{\pi_i \, q_{ij}} > 1. \end{cases} \tag{16.1}$$

Similarly, $\beta_{ji} = \pi_j \, p_{ji} = \pi_j \, q_{ji} \, \min\left\{\frac{\pi_i q_{ij}}{\pi_j q_{ji}}, 1\right\}$, so

$$\beta_{ji} = \begin{cases} \pi_j \, q_{ji} & \text{if} \quad \frac{\pi_i \, q_{ij}}{\pi_j \, q_{ji}} > 1 \\ \pi_i \, q_{ij} & \text{if} \quad \frac{\pi_i \, q_{ij}}{\pi_j \, q_{ji}} \leqslant 1. \end{cases} \tag{16.2}$$

Upon inspection, we find that the first and second rows of (16.1) are the same as the first and second rows of (16.2), respectively. This proves that $\beta_{ij} = \beta_{ji}$ in the case $i \neq j$.

6.2   Note that for this problem the acceptance probability is

$$\alpha(x, y) = \min \left\{ \exp \left\{ -\frac{(y - 10)^2 - (x - 10)^2}{2} \right\}, 1 \right\}.$$

The following Matlab script shows a simple implementation of the random walk sampler.

```
clear all, clc % clear the workspace and the screen
N=5000; % number of steps taken by sampler
sample=zeros(N,1); % preallocate memory for the vector of sample
sigma=0.1;     % standard deviation of proposal
X=randn*sigma; % generate an initializing point

for k=1:N % repeat the Metropolis step
    Y=X+randn*sigma; % generate the proposal move
    if rand<min(exp(-.5*(Y-10)^2+.5*(X-10)^2) , 1) %acceptance step
        X=Y; % update X
    end
    sample(k)=X; %store sample
end
plot(sample), grid on
xlabel('iteration t')
ylabel('state of the chain X_t')
```

A plot of typical values of samples generated by the random walk sampler is given in Figure 16.1. Roughly, stationarity seems to have been reached at about $t = 1000$.

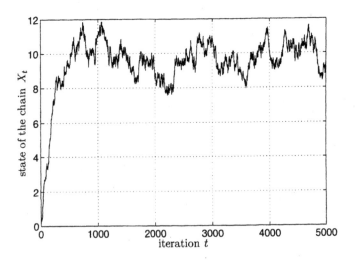

Figure 16.1 Samples generated by the random walk sampler.

6.3 The objective here is to evaluate the estimated (auto)covariance function

$$\widehat{R}(k) = \frac{1}{T-k-1} \sum_{t=1}^{T-k} (X_t - \widehat{\ell})(X_{t+k} - \widehat{\ell}), \quad k = 0, 1, \ldots, K,$$

for the stationary process in Problem 6.2. This is established by taking a *larger sample size*, say $N = 500{,}000$, in order to estimate the covariances accurately, and appending the following lines to the code in Problem 6.2 (script given for the case sigma=0.1).

```
tstat= 1000; % time process reaches stationarity
statsample = sample(tstat:N); % a stationary sample
K = 2000; % the maximum lag, should be not much bigger than tstat
[c,lags]=xcov(statsample,K,'unbiased');
plot(lags(K+1:2*K),c(K+1:2*K));
```

It is important that the samples be taken from a *stationary* process. Also, the maximum lag k for which $\widehat{R}(k)$ is estimated should not be too large. The estimate becomes unreliable for large k. In Figure 16.2 we have plotted the estimated covariance functions for three cases: $\sigma = 0.05, 0.1,$ and 0.2. The estimated times of stationarity were 3000, 1000, and 300, respectively. The maximal lags were 4000, 2000, and 600, respectively. It can be seen that choosing a smaller value for σ makes the samples highly correlated. A larger value for σ dampens the covariance function but may lead to small probabilities of acceptance. Note also that the covariance functions reach 0 around the times of stationarity.

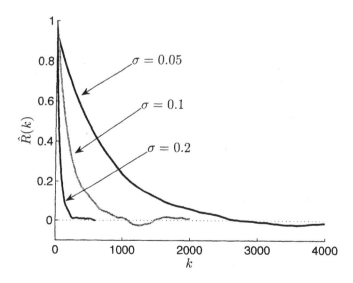

Figure 16.2 Estimated covariance function for the random walk sampler for different values of σ.

6.4 (a) For this problem the acceptance probability takes the form

$$\alpha(x,y) = \min\left\{\exp\{-(y+\lambda x)+(x+\lambda y)\}, 1\right\}\ .$$

This leads to the following Matlab script.

```
clear all, clc
N=10^5; % total number of Metropolis steps
lam=0.2; %parameter of proposal distr.: lam= 0.2, 1, 2 or 5
x=1; % initial state
sample=zeros(N,1);
for k=1:N
    y=-log(rand)/lam; % generate proposal move
    if rand<min( exp((lam-1)*y+(1-lam)*x),1)
        x=y; % accept proposal
    end
    sample(k)=x;
end
[counts,bins]=hist(sample,100); % construct a histogram
counts=counts/sum(counts)/(bins(2)-bins(1)); % normalize  histogram
bar(bins,counts), hold on % plot the normalized histogram
plot([0:0.01:10],exp(-[0:0.01:10]),'r') % superimpose the plots
```

The typical behavior for $\lambda = 0.2$ is depicted in Figure 16.3. The graph shows that there is close agreement between the true distribution and the normalized histogram of the samples obtained via the independence sampler. A similar result is obtained for values of λ in the range $[0.2, 1]$. However, when λ is larger than 2, the agreement rapidly deteriorates, as shown in Figure 16.4 for $\lambda = 5$.

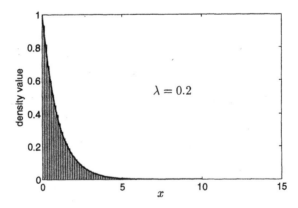

Figure 16.3 Normalized histogram versus the true density for the independence sampler with $\lambda = 0.2$.

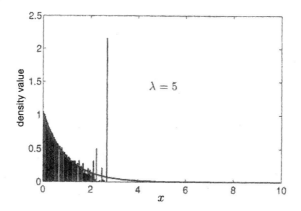

Figure 16.4 Normalized histogram versus the true density for the independence sampler with $\lambda = 5$.

(b) A typical output of the dotplots for $\lambda = 0.2, 1, 2$, and 5 is depicted in Figure 16.5. For $\lambda = 0.2$ and 1, the sample means are more or less evenly distributed about the true mean of 1, with $\lambda = 1$ giving the smallest spread. However, for $\lambda = 2$ and $\lambda = 5$ this is no longer true. The bulk of the sample means tend to fall to the left of 1, with an outlier to the right of 1 once and a while. The estimator for the mean is thus biased and tends to underestimate the true expectation.

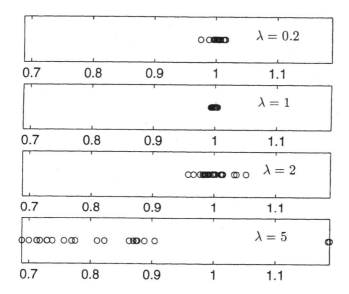

Figure 16.5 Dotplots of the sample means of twenty runs using the independence sampler with $\lambda = 0.2, 1, 2$, and 5.

For $\lambda = 1$ the samples are independent. Hence, the covariance function is 0 for lags $k = 1, 2, \dots$. For $\lambda = 0.2$ the covariance function decreases to zero in about 15 lags. For $\lambda > 2$ the covariance function cannot be accurately estimated. For example, in Figure 16.6 twenty estimates of the covariance function are given using $N = 10^6$ samples. Notice the large lags and the outlier — indicating that the true covariances are underestimated.

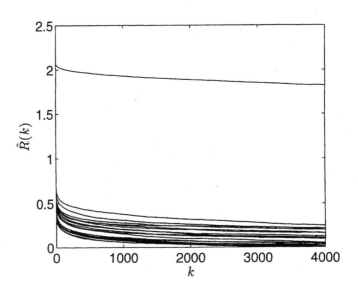

Figure 16.6   Estimated covariance functions for the independence sampler with $\lambda = 5$ for twenty runs of size $N = 10^6$.

6.5   For this problem the acceptance probability is $\alpha(x, y) = \min\left\{\exp\{x - y\} I_{\{y > 0\}}, 1\right\}$. This leads to the following script for the random walk sampler.

```
clear all,clc
N=10^5; % total number of Metropolis steps
lam=20; x=1; % initial state
sample=zeros(N,1);
for k=1:N
    rv=-log(rand)/lam; % generate from double exponential distr.
    rv=sign(rand-.5)*rv;
    Y=x+rv; % generate proposal move
    if rand<min( exp(x-Y)*(Y>0),1)
        x=Y;
    end
    sample(k)=x;
end
hist(sample,100)
```

The fact that the random walk sampler is more robust with respect to the choice of λ is illustrated in Figure 16.7, where the averages for twenty independent runs are depicted for $\lambda = 0.1, 1, 10$, and 50. In comparison to Figure 16.5 the averages are all centered around 1 with no indication of bias even for larger values of λ. Even for a value as large as $\lambda = 20$, the sample distribution is not too different from the true distribution (see Figure 16.8).

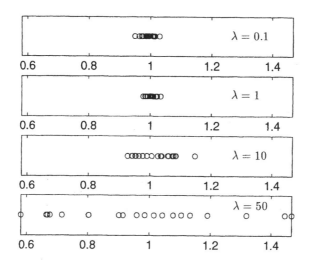

Figure 16.7 Dotplots of the sample means of twenty runs using the random walk sampler for $\lambda = 0.2, 1, 2,$ and 5.

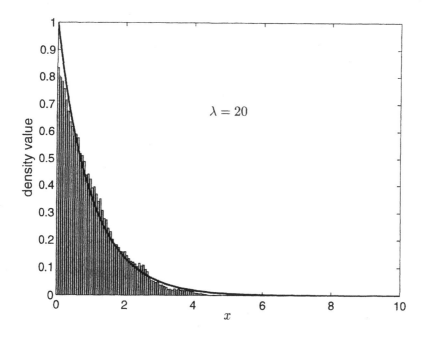

Figure 16.8 Normalized histogram versus the true density for the random walk sampler with $\lambda = 20$.

6.6 (a) The joint pdf of X and Y is given by

$$f(\mathbf{x}) \propto \exp\left\{-\frac{1}{2}\mathbf{x}^T\Sigma^{-1}\mathbf{x}\right\} = \exp\left\{-\frac{x^2 + y^2 - 2\varrho x y}{2(1 - \varrho^2)}\right\}.$$

Hence, $f(x \,|\, y) \propto \exp\left\{-\frac{(x-\varrho y)^2}{2(1-\varrho^2)}\right\}$ and $f(y \,|\, x) = \exp\left\{-\frac{(y-\varrho x)^2}{2(1-\varrho^2)}\right\}$, whence the results.

(b) A simple implementation of the systematic Gibbs sampler is the following script.

```
clear all, clc
x=0; %initializing state
y=0;
rho=.9; % correlation coefficient
sample=zeros(10^4,2);
for k=1:10^4
  x=randn*sqrt(1-rho^2)+rho*y;
  y=randn*sqrt(1-rho^2)+rho*x;

  sample(k,:)=[x,y];
end
plot(sample(:,1),sample(:,2),'r.')
```

6.7 We have

$$f_{XY}(x, y) = f_{Y|X}(y \,|\, x)f_X(x) = \frac{f_{Y|X}(y \,|\, x)f_X(x)}{\int f_Y(y)dy} = \frac{f_{Y|X}(y \,|\, x)}{\int \frac{f_Y(y)}{f_X(x)}\, dy} = \frac{f_{Y|X}(y \,|\, x)}{\int \frac{f_{Y|X}(y \,|\, x)}{f_{X|Y}(x \,|\, y)}\, dy},$$

where the last equality follows from the identity $f_X(x)f_{Y|X}(y \,|\, x) = f_Y(y)f_{X|Y}(x \,|\, y)$. The n-dimensional generalization (omitting the subscript notation) is

$$f(\mathbf{x}) \propto \prod_{i=1}^{n} \frac{f(x_i \,|\, x_1, \ldots, x_{i-1}, x_{i+1}^*, \ldots, x_n^*)}{f(x_i^* \,|\, x_1, \ldots, x_{i-1}, x_{i+1}^*, \ldots, x_n^*)},$$

where $\mathbf{x} = (x_1, \ldots, x_n)'$ and $\{x_i^*\} \in$ support of f. The proof follows by observing that

$$\prod_{i=1}^{n} \frac{f(x_i \,|\, x_1, \ldots, x_{i-1}, x_{i+1}^*, \ldots, x_n^*)}{f(x_i^* \,|\, x_1, \ldots, x_{i-1}, x_{i+1}^*, \ldots, x_n^*)}$$

$$= \prod_{i=1}^{n} \frac{f(x_i \,|\, x_1, \ldots, x_{i-1}, x_{i+1}^*, \ldots, x_n^*)f(x_{i+1}^* \cdots, x_n^*)}{f(x_i^* \,|\, x_1, \ldots, x_{i-1}, x_{i+1}^*, \ldots, x_n^*)f(x_{i+1}^* \cdots, x_n^*)}$$

$$= \prod_{i=1}^{n} \frac{f(x_1, \ldots, x_i, x_{i+1}^*, \ldots, x_n^*)}{f(x_1, \ldots, x_{i-1}, x_i^*, \ldots, x_n^*)}$$

$$= \frac{\prod_{i=1}^{n} f(x_1, \ldots, x_i, x_{i+1}^*, \ldots, x_n^*)}{\prod_{j=1}^{n} f(x_1, \ldots, x_{j-1}, x_j^*, \ldots, x_n^*)}$$

$$= \frac{f(\mathbf{x})\prod_{i=1}^{n-1} f(x_1, \ldots, x_i, x_{i+1}^*, \ldots, x_n^*)}{f(\mathbf{x}^*)\prod_{j=2}^{n} f(x_1, \ldots, x_{j-1}, x_j^*, \ldots, x_n^*)}$$

$$= \frac{f(\mathbf{x})\prod_{i=1}^{n-1} f(x_1, \ldots, x_i, x_{i+1}^*, \ldots, x_n^*)}{f(\mathbf{x}^*)\prod_{j=1}^{n-1} f(x_1, \ldots, x_j, x_{j+1}^*, \ldots, x_n^*)} = \frac{f(\mathbf{x})}{f(\mathbf{x}^*)} \propto f(\mathbf{x}).$$

The proof is valid provided that the support of $f(\mathbf{x})$ is the Cartesian product of the supports of the marginal densities $\{f(x_i)\}$.

6.8 The Ising model is a special case of the Potts model. Here, the number of sites is $J = n^2$ and each state of the system is represented via $\mathbf{x} = (x_1, \ldots, x_{n^2})$. The target density is $\pi_T(\mathbf{x}) = f(\mathbf{x}) \propto \exp\left\{\sum_{i<j} \psi_{ij}\, I_{\{x_i = x_j\}}\right\}$, where each $x_i \in \{0, 1\}$ (1 is black and 0 is white) and

$$\psi_{ij} = \begin{cases} \frac{4}{T} & \text{if } i \leftrightarrow j \\ 0 & \text{otherwise.} \end{cases}$$

The notation $i \leftrightarrow j$ means that i, j are neighboring sites. We can sample from $f(\mathbf{x})$ by introducing the auxiliary random variables Y_{ij}, $1 \leqslant i < j \leqslant n^2$ such that

$$(Y_{ij} \mid \mathbf{X} = \mathbf{x}) \sim_{\text{iid}} \mathsf{U}\left[0, \, e^{\psi_{ij}\, I_{\{x_i = x_j\}}}\right], \quad \forall i < j .$$

Hence, the conditional distribution of $\mathbf{Y} = \{Y_{ij}\}$ is

$$f(\mathbf{y} \mid \mathbf{x}) = e^{-\sum_{i<j} \psi_{ij} I_{\{x_i = x_j\}}} \prod_{i<j} I\{0 < y_{ij} < e^{\psi_{ij}\, I_{\{x_i = x_j\}}}\} .$$

Therefore,

$$f(\mathbf{y}, \mathbf{x}) = f(\mathbf{y} \mid \mathbf{x}) f(\mathbf{x}) \propto \prod_{i<j} I\{0 < y_{ij} < e^{\psi_{ij}\, I_{\{x_i = x_j\}}}\} ,$$

that is, the joint distribution is uniform on a Cartesian-product set. Hence, the conditional distribution

$$f(\mathbf{x} \mid \mathbf{y}) \propto \prod_{i<j} I\{y_{ij} < e^{\psi_{ij}\, I_{\{x_i = x_j\}}}\}$$

is also uniform. Given that sample generation from the conditional densities $f(\mathbf{x} \mid \mathbf{y})$ and $f(\mathbf{y} \mid \mathbf{x})$ is easy, we can apply the Gibbs sampler to generate a population from the joint pdf $f(\mathbf{x}, \mathbf{y})$. We can summarize the iterative procedure as follows.

Algorithm.

1. *Given a state* $\mathbf{X} = \mathbf{x}$ *for all the sites* $1 \leqslant i < j \leqslant n^2$ *that are neighbors and of the same color, draw* $Y_{ij} \sim_{\text{iid}} \mathsf{U}[0, e^{4/T}]$; *otherwise, draw* $Y_{ij} \sim_{\text{iid}} \mathsf{U}[0, 1]$.

2. *Given* $\mathbf{Y} = \mathbf{y}$, *if* $y_{ij} < 1$, *draw* X_i *and* $X_j \sim_{\text{iid}} \text{Ber}(1/2)$; *otherwise (that is, if* $1 < y_{ij} < e^{4/T}$), *draw* $X_i \sim \text{Ber}(1/2)$ *and set* $X_j = X_i$.

Setting $B_{ij} = I_{\{Y_{ij} > 1\}}$ and observing that $\mathbb{P}(B_{ij} = 1) = \mathbb{P}(Y_{ij} > 1)$, a simplified version of the algorithm is as follows.

Algorithm (Swendsen-Wang).

1. *Given a state* $\mathbf{X} = \mathbf{x}$, *generate Bernoulli random variables* $\mathbf{B} = \{B_{ij}\}$, *where*

$$B_{ij} \sim_{\text{iid}} \begin{cases} \text{Ber}(1 - e^{-4/T}) & \text{if } I_{\{x_i = x_j\}} I_{\{i \leftrightarrow j\}} = 1 \\ \text{Ber}(0) & \text{otherwise.} \end{cases}$$

2. *Given* $\mathbf{B} = \mathbf{b}$, *generate* $\mathbf{X}$ *such that*

$$\begin{cases} X_i, X_j \sim_{\text{iid}} \text{Ber}(1/2) & \text{if } b_{ij} = 0 \\ X_j \sim_{\text{iid}} \text{Ber}(1/2), \ X_i = X_j & \text{if } b_{ij} = 1 . \end{cases}$$

6.10   Let $\mathscr{X}^* = \{(x_1,\ldots,x_n), x_i \in \{0,\ldots,m\}, i = 1,\ldots,n\}$. To explain the counting procedure, suppose that $n = 4$ and $m = 3$. Some elements in $\mathscr{X}^*$ are $(0,1,1,1)$, $(3,0,0,0)$,
and $(0,0,2,1)$. Another way of looking at this is to place $m = 3$ balls into $n = 4$ urns,
numbered $1,\ldots,n$. For example, $(0,0,2,1)$ means that the third urn has two balls and
the fourth urn has one ball. One way to distribute the balls over the urns is to distribute
$n - 1 = 3$ "separators" and $m = 3$ balls over $n - 1 + m = 6$ positions, as indicated in
Figure 16.9. This can be done in $\binom{6}{3} = 20$ ways.

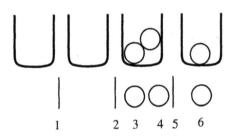

1   2 3   4 5   6

Figure 16.9   Distribute m balls over n urns.

Thus, in general, the number of ways m positions for the balls can be chosen out of
$n - 1 + m$ positions is $\binom{n+m-1}{k}$. Therefore, $|\mathscr{X}^*| = \binom{n+m-1}{m} = \binom{n+m-1}{n-1}$.

6.11 (a) We need to solve the eigenvalue problem $\mathbf{y} = \mathbf{y}P$, where $\mathbf{y} = (1, y_2, y_3)$ and

$$P = \begin{pmatrix} 1/3 & 1/3 & 1/3 \\ 1/2 & 0 & 1/2 \\ 3/4 & 1/4 & 0 \end{pmatrix} .$$

The first equation gives $\frac{1}{2} y_2 + \frac{3}{4} y_3 = \frac{2}{3}$, and the second $\frac{1}{3} + y_3 \frac{1}{4} = y_2$. Solving by
substitution yields $y_2 = \frac{10}{21}$ and $y_3 = \frac{4}{7}$.

(b) The joint pdf is

$$\pi(\mathbf{x}) = C \left(\frac{1}{2}\right)^{x_1} \left(\frac{10}{21}\right)^{x_2} \left(\frac{4}{7}\right)^{x_3} \quad \text{for } x_1 + x_2 + x_3 = 50 .$$

Therefore,

$$C^{-1} = \sum_{x_1=0}^{50} \sum_{x_2=0}^{50-x_1} \left(\frac{1}{2}\right)^{x_1} \left(\frac{10}{21}\right)^{x_2} \left(\frac{4}{7}\right)^{50-x_1-x_2} ,$$

which yields $C \approx 29664743772.8853$.

(c) To apply the Gibbs sampler in this case, we need to be able to generate random variables
easily from a truncated Geometric density. The following algorithm generates a random
variable X from the truncated density $\mathsf{TG}(p, b) \propto (1 - p)^x$, $x = 0,\ldots,b$, $0 \leqslant p \leqslant 1$
using the inverse-transform method.

Algorithm.

1. *Generate*

$$Y = \frac{\ln(1 - U[1 - (1 - p)^{b+1}])}{\ln(1 - p)}, \quad U \sim \mathsf{U}(0, 1).$$

2. *Output* $X = \lceil Y - 1 \rceil$.

This algorithm is implemented in the following m-file.

```
function rv=tgeornd(p,b)
rv=log( 1-rand(size(p)).*(1-(1-p).^(b+1)) );
rv=ceil(rv./log(1-p)-1);
```

To implement the Gibbs sampler, note that X_3 is always equal to $50 - X_1 - X_2$ and that

$$(X_2 \mid X_1) \sim \text{TG}\left(1 - \frac{7}{4} \times \frac{10}{21}, 50 - X_1\right),$$

$$(X_1 \mid X_2) \sim \text{TG}\left(1 - \frac{7}{4} \times \frac{1}{2}, 50 - X_2\right).$$

Hence, the Gibbs sampler can be used to approximately sample from $\pi(x_2, x_3)$ and thus from $\pi(x_1, x_2, x_3)$. This leads to the following algorithm.

Algorithm.

1. *Initialize with $X_1 = 1$ and then iterate Steps 2–4.*
2. *Draw $X_2 \sim \text{TG}(1/6, 50 - X_1)$.*
3. *Draw $X_1 \sim \text{TG}(1/8, 50 - X_2)$.*
4. *Set $X_3 = 50 - X_1 - X_2$.*

Once a sample of size N is obtained from $\pi(\mathbf{x})$, we can estimate C via

$$\widehat{C} = \binom{52}{2}^{-1} \frac{1}{N} \sum_{k=1}^{N} \frac{1}{\tilde{\pi}(\mathbf{X}_k)},$$

where $\mathbf{X}_k = (X_{k1}, X_{k2}, X_{k3})$ is the k-th sample and $\tilde{\pi}(\mathbf{x}) = \pi(\mathbf{x})/C$ is the unnormalized density. Using a sample size of 200,000 we obtained an estimate of $2.95 \cdot 10^{10}$, with an estimated relative error of 0.01.

12 (a) We have

$$f(\mu \mid \sigma^2, \mathbf{x}) \propto \exp\left\{-\frac{n\mu^2 - 2\mu \sum_i x_i + \sum_i x_i^2}{2\sigma^2}\right\} \propto \exp\left\{-\frac{\mu^2 - 2\mu \sum_i x_i/n}{2\sigma^2/n}\right\}$$

$$\propto \exp\left\{-\frac{(\mu - \bar{x})^2}{2\sigma^2/n}\right\}.$$

In other words, $(\mu \mid \bar{x}, \sigma^2) \sim \text{N}(\bar{x}, \frac{\sigma^2}{n})$.

(b) From the posterior pdf we see that

$$f(\sigma^2 \mid \mu, \mathbf{x}) = (2\pi)^{-n/2}(\sigma^2)^{-n/2-1} \exp\left\{-\frac{V_\mu}{2\sigma^2/n}\right\},$$

which implies that $(1/\sigma^2 \mid V_\mu)$ has a **Gamma** $(n/2, nV_\mu/2)$ distribution.

(c) In the following script we draw a random sample of size $n = 100$ from the standard normal distribution and sample from its posterior distribution using the Gibbs sampler with $N = 10^5$ samples.

```
clear all, clc
N=10^5;% Gibbs sample size
n=100; %sample size for population model
X=randn(1,n);
sample_mean=mean(X);
% initial state for Gibbs sampler
sig2=var(X);
mu=sample_mean;
A=sum(X.^2);
gibbs_sample=zeros(N,2);
for k=1:N
    mu=sample_mean+sqrt(sig2/n)*randn;
    % compute sample variance given mu
    V_m=A/n-2*mu*sample_mean+mu^2;
    % generate gamma variates and take reciprocal
    sig2=1/gamrnd(n/2,1/(n*V_m/2));
    gibbs_sample(k,:)=[mu,sig2];
end
figure(1)
hist(gibbs_sample(:,1),100) % plot the histogram of the mean,
figure(2)
hist(gibbs_sample(:,1),100) % plot the histogram of the variance
```

The results of the simulation are summarized in Figure 16.10. For this case, the sample mean and sample variance are 0.1225 and 1.1277, respectively. The solid black lines correspond to the sample mean and variance of the data (the classical estimates). The means of the posterior distributions for μ and σ^2 agree with the classical estimates for μ and σ^2, respectively.

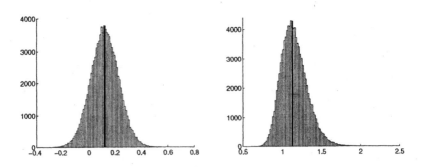

Figure 16.10   Histograms for the posterior distributions of the mean μ (left) and the variance σ^2 (right). The classical estimates for μ and σ are indicated by the solid black lines.

(d) Write

$$f(\mu \mid \mathbf{x}) = \int_0^\infty f(\mu, \sigma^2 \mid \mathbf{x}) \, d\sigma^2$$

$$= (2\pi)^{-n/2} \int_0^\infty (\sigma^2)^{-n/2-1} \exp\left\{ -\frac{V_\mu}{2\sigma^2/n} \right\} d\sigma^2$$

$$= (2\pi)^{-n/2} \int_0^\infty t^{n/2-1} \exp\left\{ -\frac{n}{2} V_\mu t \right\} dt \qquad \text{(change of var. : } t = 1/\sigma^2)$$

$$= (2\pi)^{-n/2} c^{-n/2} \, \Gamma(n/2) \int_0^\infty \frac{c^{n/2} \, t^{n/2-1} \exp\left\{ -\frac{c}{2} t \right\}}{\Gamma(n/2)} \, dt \qquad (c = n V_\mu)$$

$$= (2\pi)^{-n/2} c^{-n/2} \, \Gamma(n/2) \qquad \text{(integrand is pdf of Gamma}(n/2, c/2)) \ .$$

Therefore, substituting back $c = n V_\mu$ with $V_\mu = (\mu - \bar{x})^2 + V$, we have

$$f(\mu \mid \mathbf{x}) \propto (V_\mu)^{-n/2} = ((\bar{x} - \mu)^2 + V)^{-n/2} \propto \left(\frac{(\mu - \bar{x})^2}{V} + 1 \right)^{-n/2} .$$

6.13   The predictive pdf of Y, $f(y \mid \mathbf{x})$, where $\mathbf{x}$ are the data, can be approximated by the sample average

$$\frac{1}{N} \sum_{i=1}^N f(y \mid \Theta_i) , \qquad\qquad (16.3)$$

where the $\{\Theta_i\}$ are obtained from the posterior pdf $f(\theta \mid \mathbf{x})$ generated via the Gibbs sampler specified in Problem 6.12 (a) and (b). However, direct evaluation is computationally inefficient for large N, since for each y one has to evaluate an equally-weighted mixture of N Gaussians. Nevertheless, sampling from the equally-weighted Gaussian mixture (16.3) is not difficult. It is thus better to approximate the predictive density using a histogram constructed from a large sample from (16.3). This is accomplished by the following script.

```
clear all
x=[-0.4326, -1.6656, 0.1253, 0.2877,-1.1465]; %the data
n=length(x);
N=10^4;
sample_mean=mean(x);    % initial state for Gibbs sampler
sig2=var(x);
mu=sample_mean;

predictive_sample=zeros(N,1);
for k=1:N
    mu=sample_mean+sqrt(sig2/n)*randn;
    V_m=sum(x.^2)/n-2*mu*sample_mean+mu^2; %sample variance given mu
    sig2=1/gamrnd(n/2,1/(n*V_m/2));
    predictive_sample(k)=mu+randn*sqrt(sig2);
end
t = [-5:0.01:5];
% the variance of the classical predicitive distr.
s2p = var(x)*(1 + 1/n);
```

```
[counts,bins]=hist(predictive_sample,200); % construct a histogram
% normalize the histogram
counts=counts/sum(counts)/(bins(2)-bins(1));
bar(bins,counts)      % plot the normalized histogram
```

   The histogram is plotted in Figure 16.11. The predictive density is more spread out than the Gaussian pdf with expectation $\bar{x} = -0.566$ and standard deviation $s = 0.83$ (not plotted in the figure). In particular, the variance of the predictive sample is around 1.27. The classical approach to the predictive pdf is to use the fact that $(Y - \bar{X})/S_{\text{pred}}$, with $S_{\text{pred}} = S\sqrt{1+1/n}$, has a Student t-distribution with $n-1$ degrees of freedom. Hence, the predictive pdf of Y is approximated as

$$f_{\text{pred}}(y) = t_{n-1}\left(\frac{y-\bar{x}}{S_{\text{pred}}}\right)/S_{\text{pred}},$$

where t_{n-1} denotes the pdf of the Student t-distribution with $n-1$ degrees of freedom. Figure 16.11 shows that the classical estimate compares quite well with the Bayesian histogram.

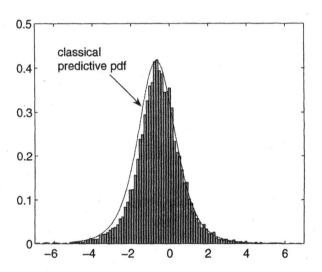

Figure 16.11 Histogram of the estimated predictive pdf and the classical predictive pdf.

6.14 (a) The joint density is

$$f(\mathbf{x}, \mathbf{r}, \lambda, p) = \frac{b^a \lambda^{a-1} e^{-b\lambda}}{\Gamma(a)} e^{-\lambda \sum_i r_i} p^{\sum_i r_i} (1-p)^{n-\sum_i r_i} \lambda^{\sum_i x_i} \prod_i \frac{r_i^{x_i}}{(x_i)!}.$$

Hence, as a function of λ, treating all other variables as constants,

$$f(\lambda \mid p, \mathbf{r}, \mathbf{x}) \propto \lambda^{a-1+\sum_i x_i} e^{-\lambda(b+\sum_i r_i)} \propto \text{Gamma}\left(a + \sum_i x_i, b + \sum_i r_i\right).$$

Similarly,

$$f(p \mid \lambda, \mathbf{r}, \mathbf{x}) \propto p^{\sum_i r_i}(1-p)^{n-\sum_i r_i} \propto \mathrm{Beta}\left(1 + \sum_i r_i, 1 + n - \sum_i r_i\right)$$

and

$$f(r_k \mid \lambda, p, \mathbf{x}) \propto \left(\frac{p\,e^{-\lambda}}{1-p}\right)^{r_k} r_k^{x_k}.$$

Hence, $f(r_k = 1 \mid \lambda, p, \mathbf{x}) \propto \frac{p\,e^{-\lambda}}{1-p}$ and $f(r_k = 0 \mid \lambda, p, \mathbf{x}) \propto I_{\{x_k=0\}}$, which implies that the distribution of r_k is given by

$$\mathrm{Ber}\left(\frac{p\,e^{-\lambda}}{p\,e^{-\lambda} + (1-p)\,I_{\{x_k=0\}}}\right).$$

(b) Using Matlab's *statistics toolbox*, this can be done via:

```
data=poissrnd(lambda,n,1).*(rand(n,1)<p);
```

(c) We chose $a = 1$ and $b = 1$ for the prior distribution of λ, and used a sample of size 10^5 obtained via the Gibbs sampler. A 95% probability interval was constructed using the following script (note that the *statistics toolbox* function gamrnd(a,b) draws from the Gamma$(a, 1/b)$ distribution).

```
clear all,clc
n=100; p=.3; lambda=2;
% generate ZIP random variables
data=poissrnd(lambda,n,1).*(rand(n,1)<p);
% now try to recover the ZIP parameters from the data
P=rand; % starting guess for p
lam=gamrnd(1,1); % starting guess for lambda
r=(rand(n,1)<P); % starting guess for  r
Sum_data=sum(data);
gibbs_sample=zeros(10^5,2);
% apply the Gibbs sampler
for k=1:10^5
    Sum_r=sum(r);
    lam=gamrnd(1+Sum_data,1/(1+Sum_r));
    P=betarnd(1+Sum_r,n+1-Sum_r);
    prob=exp(-lam)*P./(exp(-lam)*P+(1-P)*(data==0));
    r=(rand(n,1)<prob);
    gibbs_sample(k,:)=[P,lam];
end
% 95% probability interval for lambda
prctile(gibbs_sample(:,2),[2.5,97.5])
% 95% probability interval for p
prctile(gibbs_sample(:,1),[2.5,97.5])
```

The estimated Bayesian confidence intervals were found to be $(1.33, 2.58)$ for λ and $(0.185, 0.391)$ for p. Observe that the true values lie within these intervals.

6.15* Let $(x', y') \in \mathcal{R}(x, y)$ and note that $\mathcal{R}(x, y) \equiv \mathcal{R}(x', y')$, because the set of all neighbors is a communicating class. Therefore, $c(x, y) = 1 / \sum_{(m,n) \in \mathcal{R}(x,y)} f(m) \, q_m(n)$ implies that $c(x, y) = c(x', y')$. To show that μ is stationary with respect to $\mathbf{R}$, note that

$$
\begin{aligned}
\mu(x, y) \, \mathbf{R}[(x, y), (x', y')] &= \mu(x, y) \, c(x, y) \, f(x') \, q_{x'}(y') \\
&= [f(x) q_x(y)] \, c(x, y) \, f(x') \, q_{x'}(y') \\
&= [f(x') q_{x'}(y')] \, c(x, y) \, f(x) \, q_x(y) \\
&= \mu(x', y') \, c(x', y') \, f(x) \, q_x(y) \\
&= \mu(x', y') \, \mathbf{R}[(x', y'), (x, y)] \; .
\end{aligned}
$$

Therefore, the detailed balance equations (**1.38**) hold, implying the global balance equations $\mu(x, y) = \sum_{(x',y') \in \mathcal{R}(x,y)} \mu(x', y') \mathbf{R}[(x', y'), (x, y)]$, which can be written as $\mu \mathbf{R} = \mu$. We now show that for a given fixed x, $\mu(x, y)$ is stationary with respect to $\mathbf{Q}_x$. Since

$$
\begin{aligned}
\sum_{y \in \mathcal{R}(x,y')} \mu(x, y) \, \mathbf{Q}_x(y, y') &= \sum_{y \in \mathcal{R}(x,y')} f(x) \, q_x(y) \, \mathbf{Q}_x(y, y') = \\
&= f(x) \sum_{y \in \mathcal{R}(x,y')} q_x(y) \, \mathbf{Q}_x(y, y') \\
&= f(x) \, q_x(y') = \mu(x, y') \; ,
\end{aligned}
$$

the global balance equations (**1.36**) are satisfied for each $x \in \mathcal{X}$ and hence μ is stationary with respect to $\mathbf{Q}_x$. Finally, to show that μ is stationary with respect to $\mathbf{P} = \mathbf{QR}$, we use the fact that μ is stationary with respect to $\mathbf{R}$ and $\mathbf{Q}$ as follows:

$$
\begin{aligned}
&\sum_{(x,y) \in \mathcal{R}(x',y')} \mu(x, y) \, \mathbf{P}[(x, y), (x', y')] \\
&= \sum_{x \in \mathcal{R}(x',y')} \sum_{y \in \mathcal{R}(x',y')} \mu(x, y) \sum_{\tilde{y} \in \mathcal{R}(x',y')} \mathbf{Q}_x(y, \tilde{y}) \mathbf{R}[(x, \tilde{y}), (x', y')] \\
&= \sum_{\tilde{y} \in \mathcal{R}(x',y')} \sum_{x \in \mathcal{R}(x',y')} \mathbf{R}[(x, \tilde{y}), (x', y')] \sum_{y \in \mathcal{R}(x',y')} \mu(x, y) \, \mathbf{Q}_x(y, \tilde{y}) \\
&= \sum_{\tilde{y} \in \mathcal{R}(x',y')} \sum_{x \in \mathcal{R}(x',y')} \mathbf{R}[(x, \tilde{y}), (x', y')] \, \mu(x, \tilde{y}) \\
&= \sum_{(x,\tilde{y}) \in \mathcal{R}(x',y')} \mu(x, \tilde{y}) \, \mathbf{R}[(x, \tilde{y}), (x', y')] = \mu(x', y').
\end{aligned}
$$

Therefore, the global balance equations are satisfied and stationarity follows.

6.16* (a) To show that $q_x(y) = 1/n$ gives the stationary distribution of $\mathbf{Q}_x$, show that the global balance equations hold:

$$
\begin{aligned}
\sum_{y \in \mathcal{Y}} q_x(y) \, \mathbf{Q}_x(y, y') &= \sum_{y=1}^{n} \frac{1}{n} \mathbf{Q}_x(y, y') = \frac{1}{n} \left(\mathbf{Q}_x(n, y') + \sum_{y=1}^{n-1} \mathbf{Q}_x(y, y') \right) \\
&= \frac{1}{n} \left(I_{\{y'=1\}} + \sum_{y=1}^{n-1} I_{\{y+1=y'\}} \right) = \frac{1}{n} I_{\{y' \in \mathcal{Y}\}} = \frac{1}{n} = q_x(y')
\end{aligned}
$$

for all $y' \in \mathcal{Y}$.

(b) From the definition of the R-step,

$$
\begin{aligned}
\mathbf{R}[(\mathbf{x}, y), (\mathbf{x}', y)] &= c(\mathbf{x}, y) f(\mathbf{x}') q_{\mathbf{x}'}(y), \quad (\mathbf{x}', y) \in \mathcal{R}(\mathbf{x}, y) \\
&= \frac{1}{n} \frac{f(\mathbf{x}')}{[c(\mathbf{x}, y)]^{-1}} \\
&= \frac{1}{n} \frac{f(\mathbf{x}')}{\sum_{(\mathbf{z}, y') \in \mathcal{R}(\mathbf{x}, y)} f(\mathbf{z}) q_{\mathbf{z}}(y')} \\
&= \frac{f(\mathbf{x}')}{\sum_{\mathbf{z} \in \mathcal{R}(\mathbf{x}, y)} f(\mathbf{z})} .
\end{aligned}
$$

(c) On the one hand, the R-step changes only the y-th coordinate of $\mathbf{x}$ using the conditional distribution $f(\mathbf{x}') / \sum_{\mathbf{z} \in \mathcal{R}(\mathbf{x}, y)} f(\mathbf{z})$. On the other hand, the Q-step in part (a) cycles deterministically through the n components of $\mathbf{x}$, that is, $y = 1, 2 \ldots, n, 1, 2, \ldots$. This procedure is the same as the systematic Gibbs sampler.

6.17* Suppose that the target density is $f(\mathbf{x})$, $\mathbf{x} \in \mathcal{X}$ and that we wish to sample from f using the Metropolis–Hastings algorithm with transition density $q(\mathbf{y} \mid \mathbf{x}) = q(\mathbf{x}, \mathbf{y})$. We now derive the Metropolis–Hastings algorithm from the generalized Markov sampler. Consider a generalized Markov sampler over the set $\mathcal{X} \times \mathcal{Y}$, where the auxiliary set $\mathcal{Y}$ is $\mathcal{X}$. First, define the Q-step with the transition matrix

$$
\mathbf{Q}[(\mathbf{x}, \mathbf{y}), (\mathbf{x}, \mathbf{y}')] = \mathbf{Q}_{\mathbf{x}}(\cdot, \mathbf{y}') = q(\mathbf{x}, \mathbf{y}'), \quad \mathbf{y}' \in \mathcal{Y}.
$$

For a given fixed $\mathbf{x}$, the Q-step is independent of $\mathbf{y}$ and hence the stationary distribution at $(\mathbf{x}, \mathbf{y}')$ is $q(\mathbf{x}, \mathbf{y}')$ itself. Next, define the neighborhood set as $\mathcal{R}(\mathbf{x}, \mathbf{y}) = \{(\mathbf{x}, \mathbf{y}), (\mathbf{y}, \mathbf{x})\} = \mathcal{R}(\mathbf{y}, \mathbf{x})$, that is, the set has two elements only and the R-step can only exchange $\mathbf{x}$ for $\mathbf{y}$ and vice versa. More precisely, the transition probability in the R-step is $\mathbf{R}[(\mathbf{x}, \mathbf{y}), (\mathbf{y}, \mathbf{x})] = s(\mathbf{x}, \mathbf{y}) c(\mathbf{x}, \mathbf{y}) f(\mathbf{y}) q_{\mathbf{y}}(\mathbf{x})$, where $s(\mathbf{x}, \mathbf{y}) = s(\mathbf{y}, \mathbf{x})$ and the constant $c^{-1}(\mathbf{x}, \mathbf{y}) = \sum_{(\mathbf{x}', \mathbf{y}') \in \mathcal{R}(\mathbf{x}, \mathbf{y})} f(\mathbf{y}') q_{\mathbf{y}'}(\mathbf{x}') = f(\mathbf{y}) q_{\mathbf{y}}(\mathbf{x}) + f(\mathbf{x}) q_{\mathbf{x}}(\mathbf{y})$. Hence, we can write

$$
\mathbf{R}[(\mathbf{x}, \mathbf{y}), (\mathbf{y}, \mathbf{x})] = \frac{s(\mathbf{x}, \mathbf{y}) f(\mathbf{y}) q_{\mathbf{y}}(\mathbf{x})}{f(\mathbf{y}) q_{\mathbf{y}}(\mathbf{x}) + f(\mathbf{x}) q_{\mathbf{x}}(\mathbf{y})} = \frac{s(\mathbf{x}, \mathbf{y})}{1 + \varrho^{-1}},
$$

where $\varrho^{-1} = \frac{f(\mathbf{x}) q_{\mathbf{x}}(\mathbf{y})}{f(\mathbf{y}) q_{\mathbf{y}}(\mathbf{x})}$. If we now choose $s(\mathbf{y}, \mathbf{x}) = s(\mathbf{x}, \mathbf{y}) = \min\{1 + \varrho, 1 + \varrho^{-1}\}$, then the transition probability $\mathbf{R}[(\mathbf{x}, \mathbf{y}), (\mathbf{y}, \mathbf{x})]$ is equal to $\min\{\varrho, 1\}$. Therefore, given a state $\mathbf{x}$, in the Q-step a proposal state $\mathbf{y}$ is generated from $q(\mathbf{y} \mid \mathbf{x})$. The proposed $\mathbf{y}$ is then accept or rejected with probability $\alpha(\mathbf{x}, \mathbf{y}) = \min\left\{\frac{f(\mathbf{y}) q_{\mathbf{y}}(\mathbf{x})}{f(\mathbf{x}) q_{\mathbf{x}}(\mathbf{y})}, 1\right\}$ in the R-step. Thus, this particular generalized Markov sampler is equivalent to a single step of the Metropolis–Hastings algorithm.

6.18* The proof is similar to that of Problem 6.17. We take up the argument of Problem 6.17 from the point where we define the transition matrix of the R-step, that is, we are given $\alpha(\mathbf{x}, \mathbf{y}) = \mathbf{R}[(\mathbf{x}, \mathbf{y}), (\mathbf{y}, \mathbf{x})] = \frac{s(\mathbf{x}, \mathbf{y})}{1 + \varrho^{-1}}$, with $\varrho(\mathbf{x}, \mathbf{y}) = \frac{f(\mathbf{y})}{f(\mathbf{x})}$. We can now either set $s(\mathbf{x}, \mathbf{y}) = 1$ to obtain $\alpha(\mathbf{x}, \mathbf{y}) = \frac{f(\mathbf{y})}{f(\mathbf{x}) + f(\mathbf{y})}$ or define s to be a general symmetric function such that $0 \leqslant \alpha(\mathbf{x}, \mathbf{y}) \leqslant 1$.

6.19   A challenging part of the question is the programming of the objective function.

```
function s=nqueens(x,n)
% this function computes the score for an nxn chessboard,
% i.e., it counts the number of times the n-queens can capture
% ("threaten" in chess parlance) each other;
% the input is vector x  which represents the configuration
% of the queens as a permuation of [1,...,n];

% for simplicity of coding
% transform the vector x into
% matrix B. Matrix B will represent the
% chessboard: it has ones where there is a queen
%              and zero for an empty square;
for k=1:n
   B(k,x(k))=1;
end
%now compute the score
s=0; % initial score
% score cols and not rows;
a=sum(B,1);
s=s+sum((a(a>1)-1));
% score diagonal captures
a=sum(spdiags(B),1);
s=s+sum((a(a>1)-1));
a=sum(spdiags(B(n:-1:1,:)),1);
s=s+sum((a(a>1)-1));
```

Once we have a function which counts the number of times the n queens can capture each other, we can use the following script implementation of the simulated annealing with Gibbs sampling.

```
n=8; % size of chessboard
x=randperm(n); % initial state
T=1;     % initial temperature
beta=0.99; % annealing schedule
% define the function to be minimized
S=@(x)nqueens(x,n);
results=zeros(10^2,2); S_best=inf;
for t=1:10^2
   for k=1:n % apply Gibbs sampling
       for j=1:8 %compute condition density
       weights(j)=exp(-S([x(1:k-1),j,x(k+1:n)])/T);
       end
       % generate samples from the conditional density
       x(k)=randsample(8,1,true,weights);
   end
```

```
    T=T*beta; % reduce temperature
    Score=S(x);
    %keep track of the best performing state
 if Score<S_best, S_best=Score; X_best=x; end
    results(t,:)=[t,S_best];
 end
 plot(results(:,1),results(:,2),'.')
```

The typical evolution of the algorithm on this problem is illustrated in Figure 16.12.

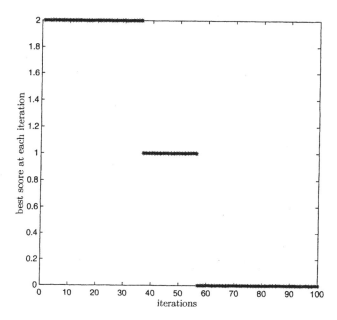

Figure 16.12 Evolution of the simulated annealing algorithm with Metropolis–Hastings sampling.

By starting the algorithm with a different random permutation, one can uncover twelve distinct solutions. Each of the twelve solutions represents an equivalence class of solutions. Within each equivalence class the solutions differ only by rotations and/or flips of the chessboard along one of its axes of symmetry. For more information see, for example, http://www.cs.rutgers.edu/~chvatal/8queens.html.

.20 The objective function (for a given cost matrix) can be computed via:

```
function s=tsp_score(x,A)
% calculate the score
     [n,m]=size(A);
s=0;
for i=1:n-1,

  s=s+A(x(i),x(i+1));
end
s=s+A(x(n),x(1));
```

As a particular example, consider the cost matrix associated with the geographical distances between 29 cities in Bavaria, given in bayg29.tsp. Note that in the script below we have assumed that the cost matrix is read from the bayg29.tsp file and is already loaded in the Matlab workspace under the name C. The best known solution (length of shortest path) is 1610 km. The typical evolution of the simulated annealing algorithm for this problem is depicted in Figure 16.13.

```
% load the cost matrix C
n=length(C); % number of nodes
x=randperm(n); %inital tour
T=1;%initial temparature
beta=0.9999;% annealing parameter
% define score function, with cost matrix C supplied by user
S=@(x)tsp_score(x,C);
results=nan(10^4,2); S_best=inf;
for t=1:10^4
    S_x=S(x);
    indices=randperm(n); % generate indices
    I=sort(indices(1:2));
    % generate proposal move
    y=[x(1:I(1)-1),x(I(2):-1:I(1)),x(I(2)+1:end)];
    S_y=S(y);
    alpha=min(exp(-(S_y-S_x)/T),1); % acceptance prob
    if rand<alpha, x=y; end
if S_best>min(S_x,S_y), S_best=min(S_x,S_y);
    X_best=x;
end
 T=beta*T;
    results(t,:)=[t,S_best];
end
plot(results(:,1),results(:,2),'.')
```

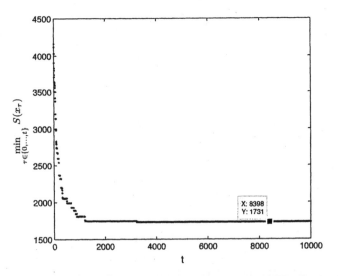

Figure 16.13   Evolution of the simulated annealing algorithm with Metropolis–Hastings sampling.

6.21   The global maximum of $S(x)$ is approximately $1.728 \approx S(1.092)$ (see Figure 16.14).

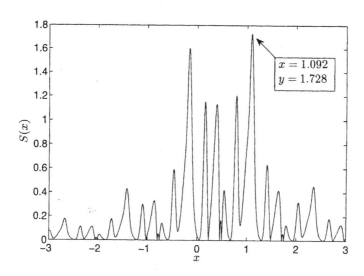

Figure 16.14   Plot of the function $S(x)$ and its maximum

To maximize the function using the simulated annealing algorithm, consider the Boltzmann target $f_t(x) = e^{S(x)/T_t}$, $t = 0, 1, \ldots$ for some annealing schedule $T_{t+1} = \beta T_t$, $0 < \beta < 1$, with $T_0 = 1$. The following script implements the idea.

```
clear all, clc
%define the function S(x)
S=@(x)abs((sin(10*x).^8+cos(5*x+1).^5)./(x.^2-x+1));
x=0; %initial state of chain
T=1; %initial temperature
beta=0.9999; %annealing parameter
sigma=.1; %standard deviation of proposal density
results=zeros(10^3,2); S_best=0;
for t=1:10^3
    %generate proposal move
    y=x+randn*sigma;
     S_x=S(x);
     S_y=S(y);
%compute the acceptance probability:
alpha=min(exp((S_y-S_x)/T),1);
%apply the Metropolis-Hastings criterion
 if  rand<alpha
       x=y; %accept proposal
 end
T=beta*T;  %reduce the temperature
% store the best function value
```

```
if max(S_x,S_y)>S_best, S_best=max(S_x,S_y); end
results(t,:)=[t,S_best];
end
plot(results(:,1),results(:,2),'.')
```

Figure 16.15 shows an output of the algorithm for values of $\sigma = 0.1$ and $\beta = 0.9999$. For such values of the tuning parameters the algorithm frequently locates the global maximum. In general, it is quite difficult to arrive at a set of parameter values that will consistently help to locate the global maximum.

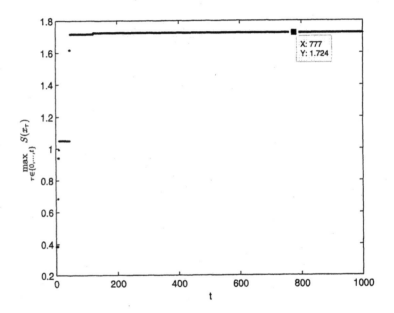

Figure 16.15   Output of the algorithm for $\sigma = 0.1$ and $\beta = 0.9999$.

CHAPTER 17

SENSITIVITY ANALYSIS AND MONTE CARLO OPTIMIZATION

7.1 (a) Let $f(\cdot; v)$ denote the density of the $\mathrm{Ber}(v)$ distribution. We have

$$
\begin{aligned}
\nabla \ell(u) &= -\frac{b}{u^2} + \frac{\partial}{\partial u} \mathbb{E}_u[X] = -\frac{b}{u^2} + \mathbb{E}_u[X\,S(u; X)] \\
&= -\frac{b}{u^2} + \mathbb{E}_u\left[\frac{X^2}{u} + \frac{X^2 - X}{1 - u}\right] \\
&= -\frac{b}{u^2} + \frac{\mathbb{E}_u[(1 - u)X^2 + u(X^2 - X)]}{u(1 - u)} \\
&= -\frac{b}{u^2} + \frac{1}{u}\mathbb{E}_u[X] \\
&= -\frac{b}{u^2} + \frac{1}{v}\mathbb{E}_v[X] \,,
\end{aligned}
$$

where we used the fact that $X^2 = X$. The corresponding likelihood ratio estimator for the gradient is

$$
\widehat{\nabla \ell}(u; v) = -\frac{b}{u^2} + \frac{1}{vN}\sum_{i=1}^{N} X_i \,,
$$

where $\{X_i\} \sim \mathrm{Ber}(v)$.

(b) The solution of the stochastic counterpart is $\widehat{u}^* = \sqrt{b\,v/\overline{X}}$, which after substitution yields $\widehat{u}^* = \sqrt{b/1.1}$.

7.2   If $f(\cdot; u)$ denotes the $\text{Exp}(u)$ density, then

$$\nabla \ell(u) = b + \mathbb{E}_v \left[X \frac{f(X;u)}{f(X;v)} \frac{\partial}{\partial u} \ln(f(X;u)) \right] = b + \mathbb{E}_v \left[X \frac{u\,e^{-uX}}{v\,e^{-vX}} \frac{1 - uX}{u} \right]$$

$$= b + \mathbb{E}_v \left[X \frac{e^{-uX}(1 - uX)}{v\,e^{-vX}} \right] .$$

Hence, the corresponding likelihood ratio estimator is

$$\widehat{\nabla \ell}(u; v) = b + \frac{1}{vN} \sum_{i=1}^{N} X_i \frac{e^{-uX_i}(1 - uX_i)}{v\,e^{-vX_i}}, \qquad \{X_i\} \sim \text{Exp}(v).$$

7.3   We have

$$\mathbb{E}_v[W^2] = \mathbb{E}_v \left[\frac{f^2(\mathbf{X}; u)}{f^2(\mathbf{X}; v)} \right] = \mathbb{E}_u \left[\frac{f(\mathbf{X}; u)}{f(\mathbf{X}; v)} \right] = \prod_{i=1}^{n} \int_0^\infty \frac{u_i^2\, e^{-2u_i\,x_i}}{v_i\,e^{-v_i\,x_i}}\, dx_i$$

$$= \prod_{i=1}^{n} \frac{u_i^2}{v_i(2u_i - v_i)} = \prod_{i=1}^{n} \frac{u_i^2}{u_i^2 - (v_i^2 - 2v_i u_i + u_i^2)}$$

$$= \prod_{i=1}^{n} \frac{1}{1 - (v_i - u_i)^2/u_i^2} = \prod_{i=1}^{n} \frac{1}{1 - \delta_i^2} ,$$

where $\delta_k = (u_k - v_k)/u_k$.

7.4   By definition

$$\mathcal{S}^{(k)}(\mathbf{u}; \mathbf{x}) = \frac{\nabla^k f(\mathbf{x}, \mathbf{u})}{f(\mathbf{x}, \mathbf{u})} ,$$

where the gradient is with respect to $\mathbf{u}$. Therefore,

$$f(\mathbf{x}, \mathbf{u})\, \mathcal{S}^{(k)}(\mathbf{u}; \mathbf{x}) = \nabla^k f(\mathbf{x}, \mathbf{u})$$

and

$$\frac{f(\mathbf{x}, \mathbf{u})}{f(\mathbf{x}, \mathbf{v})}\, \mathcal{S}^{(k)}(\mathbf{u}; \mathbf{x}) = \frac{\nabla^k f(\mathbf{x}, \mathbf{u})}{f(\mathbf{x}, \mathbf{v})} .$$

That is,

$$W(\mathbf{x}; \mathbf{u}, \mathbf{v})\, \mathcal{S}^{(k)}(\mathbf{u}, \mathbf{x}) = \nabla^k W(\mathbf{x}; \mathbf{u}, \mathbf{v}) .$$

Thus, we can show that the estimators of the sensitivities of $\ell(\mathbf{u}; \mathbf{v})$ are the sensitivities of the estimator $\widehat{\ell}(\mathbf{u}; \mathbf{v})$ as follows.

$$\nabla^k \widehat{\ell}(\mathbf{u}; \mathbf{v}) = \nabla^k \frac{1}{N} \sum_{i=1}^{N} H(\mathbf{X}_i) W(\mathbf{X}_i; \mathbf{u}, \mathbf{v})$$

$$= \frac{1}{N} \sum_{i=1}^{N} H(\mathbf{X}_i) \nabla^k W(\mathbf{X}_i; \mathbf{u}, \mathbf{v}) = \frac{1}{N} \sum_{i=1}^{N} H(\mathbf{X}_i) \mathcal{S}^{(k)}(\mathbf{u}; \mathbf{x}) W(\mathbf{X}_i; \mathbf{u}, \mathbf{v})$$

$$= \widehat{\nabla^k \ell}(\mathbf{u}; \mathbf{v}) .$$

7.5 From the independence assumption, it follows that

$$\mathbb{E}_\mathbf{v}[W^2] = \prod_{i=1}^n \mathbb{E}_{v_i}\left[\frac{\exp\{-(X-u_i)^2/\sigma_i^2\}}{\exp\{-(X-v_i)^2/\sigma_i^2\}}\right]$$

$$= \prod_{i=1}^n \int_{-\infty}^\infty \frac{1}{\sqrt{2\pi}\sigma_i} \exp\left\{-\frac{2(x-u_i)^2 - (x-v_i)^2}{2\sigma_i^2}\right\} dx$$

$$= \prod_{i=1}^n \int_{-\infty}^\infty \frac{1}{\sqrt{2\pi}\sigma_i} \exp\left\{-\frac{(x-(2u_i-v_i))^2 - 2(u_i-v_i)^2}{2\sigma_i^2}\right\} dx$$

$$= \prod_{i=1}^n \exp\left\{\frac{(u_i-v_i)^2}{\sigma_i^2}\right\}$$

$$= \exp\left\{\sum_{i=1}^n \frac{(u_i-v_i)^2}{\sigma_i^2}\right\}.$$

We also have

$$[S^{(1)}(\mathbf{u};\mathbf{X})]_i = \frac{\partial}{\partial u_i} \ln f(\mathbf{X};\mathbf{u}) = \frac{\partial}{\partial u_i} \sum_{i=1}^n \frac{(X_i-u_i)^2}{-2\sigma_i^2} = \frac{X_i-u_i}{\sigma_i^2}.$$

7.6 Since

$$\text{Var}_\mathbf{v}[H(\mathbf{X})W(\mathbf{X};\mathbf{u},\mathbf{v})] = \mathbb{E}_\mathbf{v}[H^2 W^2] - \ell^2(\mathbf{u}),$$

we have to show that $\mathbb{E}_\mathbf{v}[H^2 W^2] = \mathbb{E}_\mathbf{v}[W^2]\,\mathbb{E}_\mathbf{w}[H^2]$. This is demonstrated as follows.

$$\mathbb{E}_\mathbf{v}[H^2 W^2] = \mathbb{E}_\mathbf{v}\left[H^2(\mathbf{X})\frac{c^2(\mathbf{u})}{c^2(\mathbf{v})} \exp\left(2\sum_{i=1}^n t_i(X_i)(b_i(\mathbf{u})-b_i(\mathbf{v}))\right)\right]$$

$$= \frac{c^2(\mathbf{u})}{c(\mathbf{v})} \int H^2(\mathbf{x}) \exp\left(\sum_{i=1}^n t_i(x_i)(2b_i(\mathbf{u})-b_i(\mathbf{v}))\right) h(\mathbf{x})\, d\mathbf{x}$$

$$= \frac{c^2(\mathbf{u})}{c(\mathbf{v})\,c(\mathbf{w})} \int H^2(\mathbf{x})\, c(\mathbf{w}) \exp\left(\sum_{i=1}^n t_i(x_i)b_i(\mathbf{w})\right) h(\mathbf{x})\, d\mathbf{x}$$

$$= \frac{c^2(\mathbf{u})}{c(\mathbf{v})\,c(\mathbf{w})} \mathbb{E}_\mathbf{w}[H^2(\mathbf{X})].$$

Note that substituting for $H(\mathbf{x}) = 1$ in the last equation leads to

$$\mathbb{E}_\mathbf{v}[1^2 W^2] = \mathbb{E}_\mathbf{v}[W^2] = \frac{c^2(\mathbf{u})}{c(\mathbf{v})\,c(\mathbf{w})}\mathbb{E}_\mathbf{w}[1^2] = \frac{c^2(\mathbf{u})}{c(\mathbf{v})\,c(\mathbf{w})}1 = \frac{c^2(\mathbf{u})}{c(\mathbf{v})\,c(\mathbf{w})}.$$

Therefore, $\mathbb{E}_\mathbf{v}[W^2] = \frac{c^2(\mathbf{u})}{c(\mathbf{v})\,c(\mathbf{w})}$, and $\mathbb{E}_\mathbf{v}[H^2 W^2] = \mathbb{E}_\mathbf{v}[W^2]\,\mathbb{E}_\mathbf{w}[H^2]$, as required.

7.7 Since $\ell(u) = \int_0^\infty u\, e^{-ux} H(x)\, dx$, differentiation under the integral sign gives

$$\frac{d\ell}{du} = \int_0^\infty \left(e^{-ux} - xu\, e^{-ux}\right) H(x)\, dx = \int_0^\infty e^{-ux} H(x)\, dx - \int_0^\infty xH(x)u\, e^{-ux}\, dx.$$

Using integration by parts, the last integral is equal to

$$\int_0^\infty e^{-ux}(xH(x))'\, dx = \int_0^\infty e^{-ux} H(x)\, dx + \int_0^\infty e^{-ux} xH'(x)\, dx.$$

It follows that

$$\frac{d\ell}{du} = -\int_0^\infty e^{-ux} x H'(x)\, dx < 0\,,$$

since $H'(x) > 0$ for all x, and $u > 0$. That is, $\ell(u)$ is a monotonically decreasing function. Moreover, the second derivative is positive:

$$\frac{d^2\ell}{du^2} = -\frac{d}{du}\int_0^\infty e^{-ux} x H'(x)\, dx = \int_0^\infty e^{-ux} x^2 H'(x)\, dx > 0\,.$$

Hence, $\ell(u)$ is strictly convex.

7.8 (a) We are given the function $\mathcal{L} = \mathcal{L}(v)$ that satisfies

$$\mathcal{L}(v) = \frac{1}{\sqrt{2\pi}\sigma} \int_{-\infty}^\infty H^2(x) \exp\left\{ -\frac{(x-u)^2}{\sigma^2} + \frac{(x-v)^2}{2\sigma^2} \right\} dx$$

$$= \exp\left\{ \frac{(u-v)^2}{\sigma^2} \right\} \frac{1}{\sqrt{2\pi}\sigma} \int_{-\infty}^\infty H^2(x) \exp\left\{ -\frac{(x-(2u-v))^2}{2\sigma^2} \right\} dx$$

$$= \exp\left\{ \frac{(u-v)^2}{\sigma^2} \right\} \mathbb{E}_{2u-v}[H^2(X)] < \infty \quad \text{for } u, v < \infty\,.$$

Differentiation under the integral sign k times is justified if:

1. $H^2(x) \exp\left\{ -\frac{(x-(2u-v))^2}{2\sigma^2} \right\}$ is a Lebesgue-measurable function of x for each $v \in \mathbb{R}$.

2. The k-th order derivative $\frac{d^k}{dv^k} \exp\left\{ -\frac{(x-(2u-v))^2}{2\sigma^2} \right\}$ exists for all v and x.

3. There exists an integrable function $\Theta(x)$ such that

$$\left| H^2(x) \frac{d^k}{dv^k} \exp\left\{ -\frac{(x-(2u-v))^2}{2\sigma^2} \right\} \right| \leqslant \Theta(x) \quad \text{for all } x \text{ and } v\,.$$

Conditions 1 and 2 are obviously satisfied. We assume H is such that Condition 3 is satisfied as well. Then,

$$\mathcal{L}'(v) = \frac{1}{\sqrt{2\pi}\sigma} \int_{-\infty}^\infty H^2(x) \exp\left\{ -\frac{(x-u)^2}{\sigma^2} + \frac{(x-v)^2}{2\sigma^2} \right\} \frac{v-x}{\sigma^2}\, dx\,.$$

Also,

$$\mathcal{L}''(v) = \frac{1}{\sqrt{2\pi}\sigma} \int_{-\infty}^\infty H^2(x) \exp\left\{ -\frac{(x-u)^2}{\sigma^2} + \frac{(x-v)^2}{2\sigma^2} \right\} \left(\frac{(v-x)^2}{\sigma^4} + \frac{1}{\sigma^2} \right) dx$$

is nonnegative. Therefore, $\mathcal{L}$ is a convex function and $\mathcal{L}'$ is nondecreasing. Moreover, $\mathcal{L}$ is continuous in v, because the integrand is continuous in v and one can differentiate under the integral sign. It follows that

$$\lim_{v \to v_o} \mathcal{L}(v) = \frac{1}{\sqrt{2\pi}\sigma} \int_{-\infty}^\infty \lim_{v \to v_o} H^2(x) \exp\left\{ -\frac{(x-u)^2}{\sigma^2} + \frac{(x-v)^2}{2\sigma^2} \right\} dx = \mathcal{L}(v_o)\,.$$

Furthermore, if $\mathbb{E}_u[H^2(X)] \neq 0$ for any u, then

$$\mathcal{L}''(v) = \frac{1}{\sqrt{2\pi}\sigma} \int_{-\infty}^\infty H^2(x) \exp\left\{ -\frac{(x-u)^2}{\sigma^2} + \frac{(x-v)^2}{2\sigma^2} \right\} \frac{(v-x)^2}{\sigma^4}\, dx + \frac{1}{\sigma^2}\mathcal{L}(v)$$

$$\geqslant \frac{1}{\sigma^2}\mathcal{L}(v) = \frac{1}{\sigma^2} \exp\left\{ \frac{(u-v)^2}{\sigma^2} \right\} \mathbb{E}_{2u-v}[H^2(X)] > 0 \quad \text{for any } (2u-v) \in \mathbb{R}\,.$$

Therefore, $\mathcal{L}(v)$ is strictly convex and there exists a unique minimizer v^* of $\mathcal{L}(v)$.

(b) The derivative of $\mathcal{L}$ at $v = u$ satisfies

$$\mathcal{L}'(u) = \frac{1}{\sqrt{2\pi}\sigma} \int_{-\infty}^{\infty} H^2(x) \exp\left\{ -\frac{(x-u)^2}{\sigma^2} + \frac{(x-u)^2}{2\sigma^2} \right\} \frac{u-x}{\sigma^2} \, dx$$

$$= \frac{1}{\sqrt{2\pi}\sigma} \int_{-\infty}^{\infty} H^2(\sigma y + u) e^{-\frac{1}{2}y^2} (-y) \, dy$$

$$= \frac{-1}{\sqrt{2\pi}\sigma} \int_{-\infty}^{\infty} e^{-\frac{1}{2}y^2} \frac{d}{dy} \left(H^2(\sigma y + u) \right) \, dy < 0 \,,$$

since $H^2(x)$ is increasing. Recalling that $\mathcal{L}'$ is an increasing function and

$$\mathcal{L}'(u) < 0 = \mathcal{L}'(v^*) \,,$$

it follows that $u < v^*$.

7.9 (a) We have

$$\mathbb{E}_\eta[W^2] = \mathbb{E}_\theta[W] = \frac{1}{\sqrt{2\pi\eta\sigma^2}} \int_{-\infty}^{\infty} e^{-\frac{(x-u)^2}{2\sigma^2}} e^{-\frac{(x-u)^2}{2\sigma^2} + \frac{(x-u)^2}{2\eta^{-1}}} \, dx$$

$$= \frac{1}{\sqrt{2\pi\eta\sigma^2}} \int_{-\infty}^{\infty} \exp\left\{ -\frac{(x-u)^2}{2\frac{\eta^{-1}\sigma^2}{2\eta^{-1}-\sigma^2}} \right\} \, dx$$

$$= \frac{1}{\sqrt{2\pi\eta\sigma^2}} \sqrt{2\pi} \frac{\eta^{-1/2}\sigma}{\sqrt{2\eta^{-1}-\sigma^2}}$$

$$= \frac{1}{\sigma\sqrt{\eta}\sqrt{2-\sigma^2\eta}} = \frac{\theta}{\sqrt{\eta}\sqrt{2\theta-\eta}} \,,$$

where $\theta = \sigma^{-2}$ and $2\theta > \eta > 0$.

(b) Since $N(u, \sigma^2)$ can be reparameterized using the exponential family model $f(x; \theta) = c(\theta)e^{\theta t(x)}$, where $t(x) = -\frac{(x-u)^2}{2}$, $\theta = \sigma^{-2}$, and $c(\theta) = \sqrt{\theta/2\pi}$, we can write

$$\mathcal{L}(\eta) = \mathbb{E}_\theta[H^2 W] = \int_{-\infty}^{\infty} H^2(x) \frac{c^2(\theta)}{c(\eta)} e^{(2\theta-\eta)t(x)} \, dx$$

$$= c^2(\theta) \iint_{\mathbb{R}^2} H^2(x) e^{2\theta t(x)+\eta(t(y)-t(x))} \, dx \, dy \,.$$

It follows that

$$\mathcal{L}''(\eta) = c^2(\theta) \iint_{\mathbb{R}^2} H^2(x) e^{2\theta t(x)+\eta(t(y)-t(x))} (t(x)-t(y))^2 \, dx \, dy$$

$$\geqslant c^2(\theta) \iint_{\mathbb{R}^2} H^2(x) e^{2\theta t(x)+\eta(t(y)-t(x))} \, dx \, dy = \mathcal{L}(\eta) > 0$$

for $\eta \in (0, 2\theta)$ and any sequence $\{\eta_n\}$ that approaches 0 or 2θ. Therefore, $\mathcal{L}(\eta)$ is strictly convex and the minimizer of $\mathcal{L}(\eta)$ is unique. Note that

$$\mathcal{L}(\eta) = \frac{c^2(\theta)}{c(\eta)c(2\theta-\eta)} \mathbb{E}_{2\theta-\eta}[H^2(X)] < \infty$$

for all $\eta \in (0, 2\theta)$, because $\mathbb{E}_\eta[H^2] < \infty$ for $\eta \in (0, 2\theta)$.

(c) The derivative of $\mathcal{L}$ satisfies

$$\mathcal{L}'(\eta) = \frac{\theta}{\sqrt{2\pi}} \int_{-\infty}^{\infty} H^2(x) \frac{\partial}{\partial \eta} \left(\eta^{-1/2} e^{(2\theta-\eta)\frac{-(x-u)^2}{2}} \right) dx .$$

Direct verification shows that

$$\frac{d}{d\eta} \left(\eta^{-1/2} e^{(2\theta-\eta)\frac{-(x-u)^2}{2}} \right) \Big|_{\eta=\theta} = \theta^{-1/2} \frac{\partial}{\partial x} \left(\frac{u-x}{2\theta} e^{-\theta(x-u)^2/2} \right) .$$

Hence, by repeated use of integration by parts we find that

$$\begin{aligned}
\mathcal{L}'(\theta) &= \sqrt{\theta/(2\pi)} \int_{-\infty}^{\infty} H^2(x) \frac{\partial}{\partial x} \left(\frac{u-x}{2\theta} e^{-\theta(x-u)^2/2} \right) dx \\
&= \sqrt{\theta/(2\pi)} \int_{-\infty}^{\infty} \frac{d}{dx} (H^2(x)) \frac{x-u}{2\theta} e^{-\theta(x-u)^2/2} dx \\
&= \sqrt{\theta/(2\pi)} \int_{-\infty}^{\infty} \frac{d}{dx} (H^2(x+u)) \frac{x}{2\theta} e^{-\theta x^2/2} dx \\
&= \frac{-1}{2\theta^{3/2}\sqrt{2\pi}} \int_0^1 \frac{d}{dx} (H^2(x+u)) \frac{d}{dx} \left(e^{-\theta x^2/2} \right) dx \\
&= \frac{1}{2\theta^{3/2}\sqrt{2\pi}} \int_0^1 \frac{d^2}{dx^2} (H^2(x+u)) e^{-\theta x^2/2} dx > 0 ,
\end{aligned}$$

since H^2 is convex. We also assume that $e^{-x^2} \frac{d}{dx} H^2(x) \to 0$ as $|x| \to \infty$. Therefore, $\mathcal{L}'(\theta) > \mathcal{L}'(\eta^*) = 0$ implies $\theta > \eta^*$.

7.10   Suppose we are given that $X \sim f_X(x)$. Then, it suffices to show that $\widetilde{X} = \max(X/u, 1)$ has a density which is differentiable with respect to u, where without loss of generality we assume $u > 0$. Obviously, $\mathbb{P}(\widetilde{X} < 1) = 0$, and $\mathbb{P}(\widetilde{X} = 1) = \mathbb{P}(X/u < 1) = \int_{-\infty}^u f_X(x) \, dx$. Moreover, $\mathbb{P}(1 < \widetilde{X} < x) = \mathbb{P}(1 < X/u < x) = \int_u^{xu} f_X(z) \, dz$ for $x > 1$. Therefore, $f_{\widetilde{X}}(x) = f_X(ux) u$, where the density exists only for $x > 1$ and is differentiable with respect to u.

7.11   Note that $\mathbb{E}[\mathbf{Y}] = \mathbb{E}[\mathbf{g}(\mathbf{X})] \approx \mathbb{E}[\mathbf{g}(\boldsymbol{\mu}) + J_{\boldsymbol{\mu}}(\mathbf{g})(\mathbf{X} - \boldsymbol{\mu})] = \mathbf{g}(\boldsymbol{\mu})$. Therefore,

$$\begin{aligned}
\Sigma_{\mathbf{Y}} &= (\mathbf{Y} - \mathbb{E}[\mathbf{Y}])(\mathbf{Y} - \mathbb{E}[\mathbf{Y}])^T \\
&\approx (\mathbf{Y} - \mathbf{g}(\boldsymbol{\mu}))(\mathbf{Y} - \mathbf{g}(\boldsymbol{\mu}))^T \\
&\approx J_{\boldsymbol{\mu}}(\mathbf{g})(\mathbf{X} - \boldsymbol{\mu})[J_{\boldsymbol{\mu}}(\mathbf{g})(\mathbf{X} - \boldsymbol{\mu})]^T \\
&= J_{\boldsymbol{\mu}}(\mathbf{g})(\mathbf{X} - \boldsymbol{\mu})(\mathbf{X} - \boldsymbol{\mu})^T J_{\boldsymbol{\mu}}(\mathbf{g})^T \\
&= J_{\boldsymbol{\mu}}(\mathbf{g})\Sigma_{\mathbf{X}} J_{\boldsymbol{\mu}}(\mathbf{g})^T .
\end{aligned}$$

CHAPTER 18

THE CROSS-ENTROPY METHOD

8.1   In this case the deterministic updating formula (**8.8**) gives

$$v^* = \frac{\mathbb{E}_u[I_{\{X \geqslant 32\}} X]}{\mathbb{E}_u[I_{\{X \geqslant 32\}}]} = \mathbb{E}_u[X \mid X \geqslant 32] = 32 + u = 33 \,,$$

where the second-last equality follows directly from the memoryless property of the exponential distribution.

8.2   The Matlab script below reproduces **Table 8.1**. A typical output yields $\hat{\ell} = 1.26 \cdot 10^{-14}$, with an estimated relative error of 0.0065. The 95% confidence interval is $(1.24 \cdot 10^{-14}, 1.28 \cdot 10^{-14})$. Note that the true ℓ, which is equal to $e^{-32} \approx 1.266 \cdot 10^{-14}$, lies in this confidence interval.

```
clear all,clc, format short g
N=1000;
rho=0.05; % rarity parameter
N1=10^6;    % sample used in final (estimation) step
Gamma=32;
ell=exp(-Gamma); % exact value of rare event
v=1; % reference parameter
exit_flag=0; table=[];
for t=1:1000
    X=-log(rand(N,1))*v;
```

```
    gam=prctile(X,100*(1-rho));
    if gam>=Gamma
        gam=Gamma;    exit_flag=1;
    end
    I=(X>gam);
    W=exp(-X(I)*(1-1/v)); % form LR
    v=sum(X(I).*W)/sum(W);% update reference parameter
    table=[table;t,gam,v ];
    if exit_flag,    break,    end
end

table %display the table
% here comes the final estimation step using v*
X1=-log(rand(N1,1))*v;
I=(X1>Gamma);
W=exp(-X1*(1-1/v))*v; % form  the LR
ell_hat=mean(W.*I) % estimate ell
rel_err=std(W.*I)/sqrt(N1)/ell_hat % estimate relative  error
CI=ell_hat*[(1-1.96*rel_err), (1+1.96*rel_err)]; % compute a 95% CI
% check if true ell lies in the CI; (1 if yes)
(ell<CI(2))&(ell>CI(1))
```

8.3   The estimator is

$$\widehat{\ell} = \frac{1}{N_1} \sum_{i=1}^{N_1} I_{\{X_i>32\}} \, e^{-X_i(1-1/v)} \, v, \quad \text{with } \{X_i\} \sim_{iid} \text{Exp}(v) \,,$$

where v is any reference parameter. Hence, its variance is

$$\begin{aligned} \text{Var}(\widehat{\ell}) &= \frac{1}{N_1} \left(\mathbb{E}_v[I_{\{X>32\}} \, e^{-2X_i(1-1/v)} \, v^2] - \ell^2 \right) \\ &= \frac{1}{N_1} \left(\frac{v^2}{2v-1} \mathbb{E}_{\frac{v}{2v-1}} [I_{\{X>32\}}] - \ell^2 \right) \\ &= \frac{1}{N_1} \left(\frac{v^2}{2v-1} e^{-32(2-1/v)} - \ell^2 \right) \,. \end{aligned}$$

Therefore, the true relative error is

$$\text{RE} = \sqrt{\text{Var}(\widehat{\ell})}/\ell = N_1^{-1/2} \sqrt{\frac{v^2}{2v-1} e^{32/v} - 1} \,. \tag{18.1}$$

For this particular problem, the optimal reference parameter is $v^* = 32 + 1 = 33$, which gives a true relative error of RE = 0.0065713 (rounded). This is close to the estimated relative error of 0.0065. The relative error of the CMC estimator, for a given sample size N, can be computed by substituting $v = 1$ in (18.1), giving $8.8861 \cdot 10^6 N^{-1/2}$ (rounded). Therefore, to achieve a relative error of 0.0065713 with CMC, we need approximately

$$N = \left(\frac{8.8861 \cdot 10^6}{0.0065713} \right)^2 \approx 1.8286 \cdot 10^{18} \quad \text{samples.}$$

8.4   The following script could be used to produce a table similar to **Table 8.3**.

```
clear all
% first estimate the optimal reference parameters
N=10^3;N1=10^5;Gamma=6;rho=0.1;
u=[1 , 1 , 0.3 ,0.2, 0.1];v=u;
table=[]; exit_flag=0;
for t=1:60
   X=-log(rand(N,5))*diag(v);
   Scores=S(X);
   gam=prctile(Scores,100*(1-rho));
   if gam>=Gamma
      gam=Gamma; exit_flag=1;
   end
   I=(Scores>gam);
   Xel=X(I,:); % elite samples
   w=exp(-Xel*(1./u-1./v)')*prod(v./u);
   w=w/sum(w); % normalize weights
   v=w'*Xel ;   % update optimal reference parameter
   table=[table;t,gam,v];
   if exit_flag,  break,  end
end
% proceed to estimate the rare event
% generate from importance sampling density
X=-log(rand(N1,5))*diag(v);
Scores=S(X); I=(Scores>Gamma);
Xel=X(I,:);
w=exp(-Xel*(1./u-1./v)')*prod(v./u);   % compute LR
ell_hat=sum(w)/N1      % estimator of rare event prob
rel_err=sqrt((sum(w.^2)/N1-ell_hat^2)/N1)/ell_hat
table        % view table
```

Here S is defined in the following m-file:

```
function out = S(X) % define the objective  function S(x)
out=min([X(:,1)+X(:,4),  X(:,2)+X(:,5),...
  X(:,1)+X(:,3)+X(:,5),  X(:,2)+X(:,3)+X(:,4)],[],2);
```

8.5   To reproduce **Table 8.4**, we can use exactly the same script as for Problem 8.4, except that
we set $u = (1, 1, 1, 1, 1)$ and the objective function is redefined as per the m-file below.

```
function out=S(Z)
alpha=2;
u=[1,1,.3,.2,.1];
X=Z.^(1/alpha)*diag(u);
out=min([X(:,1)+X(:,4),  X(:,2)+X(:,5),...
  X(:,1)+X(:,3)+X(:,5),  X(:,2)+X(:,3)+X(:,4)],[],2);
```

For $\gamma = 2$ and $\alpha = 5$, a typical evolution of the algorithm is given in the following table.

Table 18.1   The evolution of $\hat{\mathbf{v}}_t$ for estimating $\mathbf{v}^*$. The estimated probability is $\hat{\ell} = 2.7 \cdot 10^{-19}$ with an estimated relative error of 0.2.

t	$\hat{\gamma}_t$	$\hat{v}_{1t}$	$\hat{v}_{2t}$	$\hat{v}_{3t}$	$\hat{v}_{4t}$	$\hat{v}_{5t}$
0		1	1	1	1	1
1	1.1604	1.8049	2.2304	0.99298	1.2928	1.1811
2	1.3264	2.9835	3.9039	1.3435	1.3211	1.1862
3	1.4478	4.1564	5.4761	0.98486	1.3733	1.4116
4	1.5515	5.4974	7.2731	0.88115	1.5061	1.2834
5	1.6317	6.2363	9.1349	0.68806	2.2809	1.8702
6	1.6994	7.7367	11.414	0.8145	3.7893	1.5692
7	1.7915	9.8353	14.083	0.57929	2.5629	2.4509
8	1.8372	13.794	15.643	0.2188	1.1574	3.526
9	1.9137	16.167	22.514	0.24845	2.2596	4.204
10	2	20.469	24.601	0.1938	3.0873	5.2443

8.6   The script below implements the CE method with $\text{Weib}(\alpha, v_i^{-1})$ as the importance sampling distribution for the i-th component. Note that the updating formula for the i-th component at iteration t is

$$\hat{v}_{t,i} = \left(\frac{\sum_{k=1}^{N} I_{\{S(\mathbf{X}_k) > \hat{\gamma}_t\}} W(\mathbf{X}_k; \mathbf{u}, \hat{\mathbf{v}}_{t-1}) X_{ki}^{\alpha}}{\sum_{k=1}^{N} I_{\{S(\mathbf{X}_k) > \hat{\gamma}_t\}} W(\mathbf{X}_k; \mathbf{u}, \hat{\mathbf{v}}_{t-1})} \right)^{1/\alpha} .$$

```
clear all
% define the objective  function S(x)
S=@(X)min([X(:,1)+X(:,4),  X(:,2)+X(:,5),...
    X(:,1)+X(:,3)+X(:,5),  X(:,2)+X(:,3)+X(:,4)],[],2);
% define the Weibull(alpha, lambda) pdf
weib=@(x,alpha,lambda)...
(alpha.*lambda.*(lambda.*x).^(alpha-1).*exp(-(lambda.*x).^alpha));
% first estimate the optimal reference parameters
N=10^3;N1=10^5;Gamma=10^4;rho=0.1;
alpha=0.2;
u=[1 , 1 , .3 ,.2, .1]; v=u;
table=[]; exit_flag=0;
for t=1:60
  % generate weibull(alpha,1./v) r.v.'s
  X=(-log(rand(N,5))).^(1/alpha)*diag(v);
  Scores=S(X);
  gam=prctile(Scores,100*(1-rho));
  if gam>=Gamma
        gam=Gamma;    exit_flag=1;
  end
  I=(Scores>gam);
  Xel=X(I,:); % elite samples
```

```
    w=prod(weib(Xel,alpha,repmat(1./u,sum(I),1))./...
        weib(Xel,alpha,repmat(1./v,sum(I),1)),2);
    w=w/sum(w); % normalize weights
    v=(w'*(Xel.^alpha)).^(1/alpha); % updating  for weibull case
    table=[table;t,gam,v ];
    if exit_flag,    break,    end
end
%proceed to estimate the rare event
% generate from importance sampling density
X=(-log(rand(N1,5))).^(1/alpha)*diag(v);
Scores=S(X); I=(Scores>Gamma);
Xel=X(I,:);
w=prod(weib(Xel,alpha,repmat(1./u,sum(I),1))./...
    weib(Xel,alpha,repmat(1./v,sum(I),1)),2); % compute LR
ell_hat=sum(w)/N1 % estimator of rare event prob
rel_err=sqrt((sum(w.^2)/N1-ell_hat^2)/N1)/ell_hat
table % view table
```

A typical output is given in the following table.

Table 18.2   The evolution of $\widehat{v}_t$ for estimating v^*. The estimated probability is $\widehat{\ell} = 3.294 \cdot 10^{-6}$ with an estimated relative error of 0.034.

t	$\widehat{\gamma}_t$	$\widehat{v}_{1t}$	$\widehat{v}_{2t}$	$\widehat{v}_{3t}$	$\widehat{v}_{4t}$	$\widehat{v}_{5t}$
0	1	1	1	1	1	1
1	3.9694	34.812	10.697	0.79775	0.90339	0.99923
2	53.959	166.69	239.19	0.27188	1.5575	0.31149
3	336.69	783.03	968.56	0.29277	2.0988	0.48903
4	1216.3	3123	3130.5	0.086391	0.31591	0.14751
5	5597.7	12840	12995	0.073041	0.90897	0.073177
6	10000	19888	16535	0.25296	0.012822	0.31241

8.7   The following script solves the root-finding problem for a given ℓ.

```
clear all
% first estimate the optimal reference parameters
N=10^3;N1=10^5;ell=10^-5;rho=0.1;
u=[.25 , .5 , 0.1 ,0.3, 0.2];v=u;
table=[];
for t=1:600
X=-log(rand(N,5))*diag(v);
Scores=S(X); gam=prctile(Scores,100*(1-rho));
I=(Scores>gam); Xel=X(I,:); % elite samples
w=exp(-Xel*(1./u-1./v)')*prod(v./u);
hat_ell_t=max(ell,sum(w)/N);
v=w'*Xel/sum(w) ;    % update optimal reference parameter
table=[table;t,hat_ell_t,v ];
```

```
    if hat_ell_t==ell,      break,      end
end
%once we have estimated the necessary parameters,
% solve the root-finding problem many
% times to obtain a bootstrap sample of size 100
for k=1:100
gamma(k)=root_finding(ell,N1,u,v);
end
```

The script calls the objective function $S(x)$ (implemented as in the solution to Problem 8.4) and the following function root_finding.m.

```
function gamma=root_finding(ell,N1,u,v)
X=-log(rand(N1,5))*diag(v); % generate from IS density
[Scores,Index]=sort(S(X),'descend');
w=exp(-X*(1./u-1./v)')*prod(v./u); % compute LR for all X
% now find the smallest gamma for which ell_hat<=ell
ell_hat=0;k=0;
while ell_hat<=ell
    k=k+1;
    ell_hat=ell_hat+w(Index(k))/N1;
end
gamma=Scores(k-1); % an estimate of the root
```

A bootstrap sample of size 100 gave a mean of $\widehat{\gamma} = 2.1757$ with an estimated relative error of 0.000226.

8.8   The following script can be used to find an optimal cut.

```
clear all, format short g
C=zeros(20,20); % construct the cost matrix
C(1,2)=1; C(1,5)=1; C(1,16)=1; C(2,1)=1;C(2,3)=1;C(2,15)=1;
C(3,2)=1; C(3,4)=1; C(3,13)=1; C(4,3)=1; C(4,5)=1; C(4,8)=1;
C(5,4)=1; C(5,1)=1; C(5,6)=1; C(6,7)=1; C(6,5)=1; C(6,18)=1;
C(7,6)=1; C(7,8)=1; C(7,9)=1; C(8,7)=1; C(8,11)=1; C(8,4)=1;
C(9,7)=1; C(9,10)=1; C(9,19)=1; C(10,9)=1; C(10,11)=1; C(10,12)=1;
C(11,10)=1;C(11,8)=1;C(11,13)=1;C(12,10)=1;C(12,20)=1;C(12,14)=1;
C(13,3)=1;C(13,11)=1;C(13,14)=1;C(14,12)=1;C(14,13)=1;C(14,15)=1;
C(15,2)=1;C(15,17)=1;C(15,14)=1;C(16,1)=1;C(16,18)=1;C(16,17)=1;
C(17,15)=1;C(17,16)=1;C(17,20)=1;C(18,16)=1;C(18,6)=1;C(18,19)=1;
C(19,18)=1;C(19,9)=1;C(19,20)=1;C(20,19)=1;C(20,17)=1;C(20,12)=1;
rho=0.1; n=20;  N=1000; % sample size
p=[1,0.5*ones(1,n-1)]; % initial cutting probabilities
tab=[];
for iter=1:400
  X=(rand(N,n)<repmat(p,N,1)); % create N number of cut vectors
  Scores=S(X,C);
  gam=prctile(Scores,(1-rho)*100);
```

```
Index=(Scores>=gam);
p=mean(X(Index,:),1); % update based on the elite sample
p(1)=1; % keep the first element of X in one sets at all times
tab=[tab;iter,gam,max(Scores)];
if max(min(p,1-p))<10^-4 ,break, end
end
tab
```

The objective function for a given cost matrix is computed via the following m-file:

```
function perf = S(x,C)
% C is the cost matrix
% x is the cut vector
N = size(x,1);
for i=1:N
 V1 = find(x(i,:)); % {V1,V2} is the partition
 V2 = find(~x(i,:));
perf(i,1)=sum(sum(C(V1,V2))); % the size of the cut
end
```

There are a total of 250 distinct cuts. If we always assign the first component of the cut vector to be 1, then we have 125 cuts; ignoring the five-fold symmetry, we have 25 cut vectors.

8.9   A typical cost matrix is depicted in Figure 18.1. The area of the two white regions — which represents the cost associated with the cut $V = \{\{1,\ldots,m\},\{m+1,\ldots,n\}\}$ — is always $cm(n-m)$. The area of the larger non-white square is always less than or equal to $b(n-m)^2$. The cut V will be optimal if the area of the white rectangle is larger than $b(n-m)^2$, that is, if $c > b(n-m)/m$.

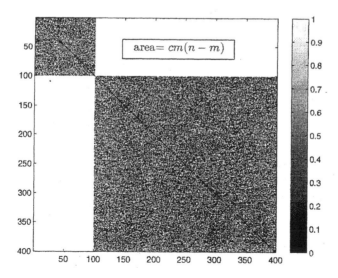

Figure 18.1   Typical synthetic cost matrix for $n = 400$, $m = 100$, $c = 1$, $a = 0$, and $b = 1$.

The following script was used to generate Table 18.3, displaying the evolution of the CE algorithm for this max-cut problem. The script calls the score function $S(x, C)$ coded in the previous problem.

```
clear all, format short g
% first create the synthetic max-cut cost matrix
m=200; n=400; a=0; b=1; c=1; % set up dimension of C and paramters
Z_1=tril(a+b*rand(m,m),-1);Z_1=Z_1+Z_1';
Z_2=tril(a+b*rand(n-m,n-m),-1);Z_2=Z_2'+Z_2;
C=[Z_1,c*ones(m,n-m);c*ones(n-m,m),Z_2]; % cost matrix
p_star=[ones(1,m),zeros(1,n-m)]; % initialize optimal cut vector
gamma_star=c*m*(n-m); %the optimal value of the cut
rho=0.1; % rarity parameter
N=1000; % sample size
p=[1,0.5*ones(1,n-1)]; % initial cutting probabilities
tab=[];
for iter=1:400 % set up iteration counter
  X=(rand(N,n)<repmat(p,N,1)); % create N number of cut vectors
  [Scores,Index]=sort(S(X,C),'descend'); X=X(Index,:);
  gam=Scores(floor(rho*N));
  p=mean(X(1:floor(rho*N),:),1);
  % keep the first element of X in one of the sets at all times
  p(1)=1;
  % difference between optimal solution and CE solution
  error=norm(p-p_star);
  tab=[tab;iter,gam,Scores(1),error]
  if error==0, break ,end
end
tab
```

Table 18.3 The evolution of the CE algorithm for the max-cut problem with $n = 400$, $m = 200$, $\varrho = 0.1$, and $N = 1,000$. Note that $x_1 \in V_1$, so that $\widehat{p}_{1,t} = 1$ for all t.

t	$\widehat{\gamma}_t$	$\max_k S(\mathbf{X}_{k,t})$	$\|\widehat{\mathbf{p}}_t - \mathbf{p}\|$
1	30063	30269	9.88
2	30070	30206	9.5122
3	30123	30350	8.8232
4	30364	30768	8.0748
5	30737	31197	7.3114
6	31237	31810	6.6478
7	31784	32748	5.9572
8	32512	33362	5.3059
9	33213	34206	4.6793
10	33989	34889	4.0548
11	34890	35701	3.3614
12	35796	37009	2.7157
13	36669	37504	2.1815
14	37474	38577	1.6693
15	38191	39018	1.2247
16	38844	39318	0.81271
17	39386	39801	0.47371
18	39795	40000	0.21307
19	40000	40000	0

.10 (a) It is straightforward to see that the tour $\mathbf{x}^* = (1, 2, \ldots, n)$ is optimal by noting that any tour must select n elements from the matrix C, one in each row. Since $a > 1$, the smallest non-diagonal n elements are the n 1s in the matrix. These form a valid tour (given by $\mathbf{x}^*$), with score $\gamma = n$; and as we have already picked up the smallest elements of C that give a valid tour, we must have $\gamma = \gamma^*$ — the minimal score over all tours.

(b) The performance of the CE algorithm applied to the synthetic TSP of size $n = 30$ is listed in Table 18.4. Here *best* and *worst* are the best and worst of the elite samples. We used the problem parameters $a = 1.5$ and $b = 3$, and the CE parameters $\alpha = 0.7$, $N = 4,500$, and $\varrho = 0.1$, giving an elite sample size of $N_e = 45$.

Table 18.4 Performance of the CE algorithm applied to the synthetic TSP of size $n = 30$.

Iter	Best	Worst	p_t^{mm}
1	57.5927	75.6859	0.0345
2	51.9153	72.3605	0.0726
3	45.9192	69.4029	0.0949
4	41.7099	63.9388	0.1374
5	37.6785	61.2835	0.2097
6	33.2758	56.7471	0.2185
7	30.0000	54.2385	0.2522
8	30.0000	47.8083	0.7299
9	30.0000	46.7860	0.9190
10	30.0000	42.2231	0.9757

.11 Table 18.5 summarizes the results for a number of benchmark problems from the TSP library.

Table 18.5 Case studies for the asymmetric TSP.

file	γ^*	min	mean	max	CPU	$\bar{T}$
br17	39	39	39	39	0.23	13.4
ftv33	1286	1286	1343.4	1372	5.0	24.4
ftv35	1473	1475	1483.7	1500	6.3	25.5
ftv38	1530	1536	1543.6	1554	9.32	27
ftv44	1613	1622	1652.2	1669	20.0	32.
ftv47	1776	1789	1820.4	1842	30.3	37.8
ft53	6905	7078	7148.2	7204	47.3	41.5
ftv55	1608	1614	1645.4	1701	58.5	42.1
ftv64	1839	1839	1849	1879	108.4	44.4
ftv70	1950	1957	1964.6	1981	185	54.3
kro124p	36230	37706	38429.8	39037	1387	110
p43	5620	5622	5624	5632	32.9	61.4
ry48p	14422	14467	14761.8	15020	33.6	41.5

.12 In all of the following experiments we consider the symmetric TSP problem ulysses16.tsp, with minimum path length 6,859. The CE algorithm was terminated when no progress was observed over three successive iterations.

(a) The left-hand graph in Figure 18.2 depicts the minima found by the CE method for a range of values of ϱ. Observe that the smaller the rarity parameter ϱ, the more likely it is to be trapped in a local minimum. The CE parameters are $N = 5n^2 = 1,280$, and $\alpha = 0.7$. The right-hand graph in Figure 18.2 shows that the number of iterations required for the algorithm to converge increases with ϱ.

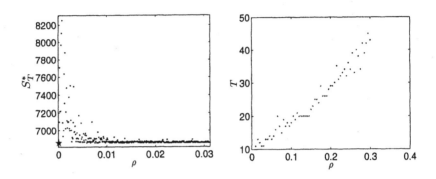

Figure 18.2 The minimum S_T^* found by the CE method (left) and the number of iterations required for the algorithm to converge (right), for a range of values of ϱ. The CE parameters are $N = 5n^2 = 1280$, and $\alpha = 0.7$. The global minimum path is designated by the star in the bottom left corner of the left figure.

The left-hand graph of Figure 18.3 shows that the smaller the smoothing parameter α, the less likely it is that algorithm is trapped in a local minimum. The right-hand graph shows that larger values of the smoothing parameter lead to faster convergence. The CE parameters are $N = 1280$ and $\varrho = 0.01$.

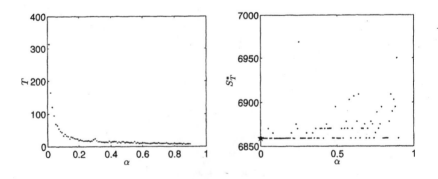

Figure 18.3 The minimum S_T^* found by the CE method (left) and the number of iterations T required by the CE method to converge (right), for a range of values of $0 < \alpha \leqslant 1$.

(b) From part (a) we know that reducing ϱ tends to trap the algorithm in a local minimum and that reducing α prevents, on average, the convergence to a local minimum. It is then reasonable to expect that that these two counteracting tendencies can be balanced in a way that avoids convergence to a local minimum. This is indeed the case and, for example, any of the set of values $\varrho_i = \alpha_i = i/100$, $i = 1, \ldots, 32$ with $N = 1{,}280$ is good enough to avoid convergence to a local minimum.

(c) Figure 18.4 demonstrates that (everything else remaining equal) the larger the sample size, the less likely the algorithm is to become trapped in a local minimum. It is then plausible that by simultaneously increasing the sample size, one can counteract the negative effect of reducing ϱ, for example. For instance, the optimal solution was found for any of the pair of values $N_j = 1000j$, $\varrho_j = 100/N_j$, $j = 1, \ldots, 10$, and $\alpha = 0.7$.

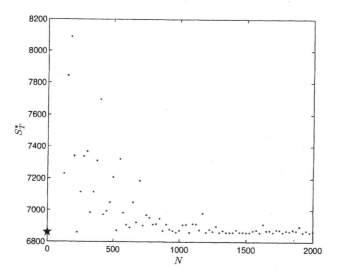

Figure 18.4 The minimum S_T^* found by the CE method for a range of samples sizes. Here we have $\varrho = 0.01$ and $\alpha = 0.7$.

.13   Running the CE algorithm 100 times on the Hammersley problem generated 36 distinct optimal tours.

14 (a) The objective function for the longest path problem is

$$S(\mathbf{x}, k) = \sum_{i=1}^{k} c_{x_i, x_{i+1}}, \quad \{x_i\}_{i \neq 1} \in \{2, \ldots, n - 1\}.$$

Similar to the TSP, x_1 is always set to 1. In addition, x_{k+1} is always set to n. The objective function accepts an additional (auxiliary) variable $k \in \{2, \ldots, n - 1\}$ that specifies the length of the path, that is, the length of a subset of a permutation on the integers from 1 to n with fixed endpoints 1 and n.

(b) The generation mechanism is as follows.

1. Generate a permutation on $2, \ldots, n-1$ according to node placement or node transition, exactly as in the TSP. (The varying length of the path is taken into account in the definition of the score function.)

2. Generate an auxiliary variable K according to a discrete distribution on $2, \ldots, n-1$, which will correspond to the length of the path.

The initial distribution of $\mathbf{X}$ is uniform over all permutations on $2, \ldots, n-1$. The updating for the permutation-generation mechanism is still the same as for the TSP, since the auxiliary variable K takes into account the varying path length.

The initial distribution of the auxiliary variable K is uniform on 2 to $n-1$, with subsequent updating of the distribution of K based on the best performing samples. The updating formula for the distribution of $\{K = i\}$ in step t is

$$\widehat{\mathbb{P}}_t(K = i) = \frac{\sum_{j=1}^{N} I_{\{S(\mathbf{X}_j, K_j) \geqslant \widehat{\gamma}_t\}} I_{\{K_j = i\}}}{\sum_{j=1}^{N} I_{\{S(\mathbf{X}_j, K_j) \geqslant \widehat{\gamma}_t\}}}, \quad i = 2, \ldots, n-1 ,$$

where $\{K_j\}$ are discrete random variables generated in step $t-1$ according to $\widehat{\mathbb{P}}_{t-1}$.

(c) Our test problem uses the 24×24 cost matrix

$$C = \begin{pmatrix}
* & * & * & 1 & * & * & * & * & * & * & * & 1 & * & 1 & 1 & 1 & * & 1 & 1 & * & * & * & * & * \\
* & * & * & * & * & * & * & * & * & 1 & * & 1 & 1 & 1 & * & * & * & * & * & * & 1 & * & * & * \\
* & * & * & 1 & * & 1 & * & * & * & * & * & 1 & * & 1 & * & * & * & * & * & * & * & * & * & * \\
1 & * & * & * & * & 1 & * & * & * & * & 1 & * & * & * & * & 1 & * & 1 & * & * & * & * & * & * \\
* & * & 1 & * & * & * & * & * & * & * & * & * & * & * & 1 & 1 & * & 1 & * & * & * & * & * & * \\
* & * & * & * & * & 1 & * & * & 1 & * & 1 & * & 1 & * & 1 & * & * & * & * & * & * & * & * & * \\
* & * & 1 & 1 & * & 1 & * & * & 1 & * & 1 & 1 & * & * & 1 & * & * & * & * & * & * & * & * & * \\
* & * & * & * & * & * & 1 & 1 & * & * & * & 1 & * & 1 & * & 1 & * & * & * & * & * & 1 & * & * \\
* & * & * & * & * & * & 1 & 1 & * & * & * & * & * & * & 1 & * & * & * & * & * & * & * & * & * \\
* & 1 & * & * & * & 1 & * & * & * & * & * & * & * & 1 & 1 & * & 1 & 1 & * & * & * & * & * & * \\
1 & * & 1 & * & * & 1 & * & * & * & 1 & * & * & * & * & * & * & * & * & * & 1 & * & * & * & * \\
* & 1 & 1 & * & * & 1 & 1 & * & * & * & 1 & * & * & * & * & * & 1 & * & 1 & * & 1 & * & * & * \\
* & 1 & * & * & * & * & 1 & * & * & * & * & * & * & * & * & * & 1 & * & 1 & * & * & * & * & * \\
1 & 1 & * & * & * & 1 & * & * & 1 & * & * & * & * & * & * & * & * & * & * & * & * & * & * & * \\
1 & * & * & * & 1 & * & * & * & 1 & * & 1 & * & * & * & * & * & * & * & * & 1 & * & * & 1 & 1 \\
1 & * & * & 1 & 1 & * & 1 & 1 & 1 & * & 1 & * & * & * & * & * & * & * & 1 & 1 & * & 1 & * & * \\
* & * & * & * & * & * & * & * & * & * & * & * & * & * & * & * & * & * & 1 & 1 & * & 1 & 1 & * \\
1 & * & * & 1 & 1 & * & * & * & * & * & 1 & 1 & * & * & 1 & * & 1 & * & * & * & * & * & * & * \\
1 & * & * & * & * & * & * & * & * & 1 & * & * & * & 1 & 1 & * & * & * & * & * & 1 & * & * & * \\
* & * & * & * & * & * & * & * & * & 1 & * & 1 & * & * & * & * & * & * & * & * & * & * & * & * \\
* & 1 & * & * & * & * & * & 1 & * & * & * & * & * & * & * & * & 1 & * & * & * & * & * & * & * \\
* & * & * & * & * & * & * & * & * & * & * & * & 1 & 1 & * & * & 1 & 1 & 1 & * & 1 & * & * & 1 \\
* & * & * & * & * & * & * & * & * & * & * & * & * & * & 1 & * & * & * & * & * & * & 1 & * & * \\
\end{pmatrix} .$$

A star denotes a cost of $-\infty$, corresponding to an impossible link in the path. Table 18.6 depicts the evolution of the CE algorithm for this longest path problem. The CE parameters were $\varrho = 0.1$, $N = 5{,}000$, and $\alpha = 0.7$. The optimal solution found in this case is given by $\mathbf{x} = (1, 20, 17, 4, 12, 13, 21, 11, 2, 14, 19, 5, 3, 7, 6, 15, 10, 8, 9, 22, 18, 23, 16, 24)$.

Table 18.6 Evolution of the CE algorithm on the longest path problem with matrix C given above.

t	S_t^*	$\widehat{\gamma}_t$
1	13	8
2	15	10
3	17	12
4	19	14
5	20	16
6	21	17
7	22	18
8	22	19
9	23	19
10	23	20
20	23	22
30	23	22
40	23	23
50	23	23

Fig-ure 18.5 illustrates the convergence of the distribution of the path length K. Note that gradually more and more weight is placed on longer and longer paths. On iteration 17 the distribution had converged to the degenerate pdf with $\mathbb{P}(K = 23) = 1$.

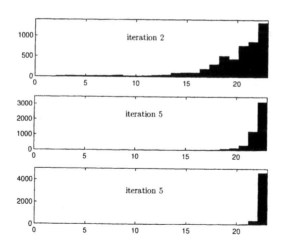

Figure 18.5 Evolution of the distribution of K across iterations.

15 The following script implements the CE method for the n-queens problem (see also the solution of Problem 6.19). The objective function nqueens.m is the same as in Problem 6.19.

```
clear all,n=8; % size of chessboard
N=500; alpha=0.7; rho=0.1;
S=@(x)nqueens(x,n); % define the n-queens score function
results=[]; S_best=inf;
P=.5*ones(n,n); % initial probabilities.
for t=1:50
    % generate sample
    for i=1:n
        x(:,i)=randsample(n,N,true,P(:,i));
    end
    % score the sample
    for k=1:N
        Score(k)=S(x(k,:));
    end
    gam=prctile(Score,rho*100);
    Index=(Score'<=gam);

    for i=1:n     % update the probabilities based on elite sample
        P(i,:)=alpha*mean(x(Index,:)==i,1)+ (1-alpha)*P(i,:);
    end
    %keep track of the best performing state
    if min(Score)<S_best, [S_best,I]=min(Score); X_best=x(I,:); end
    results=[results;t,S_best];
    % monitor progress
    pause(.01)
    plot(results(:,1),results(:,2),'.')
end
```

The evolutions of the eight-queens problem via the CE method and the simulated annealing algorithm are similar.

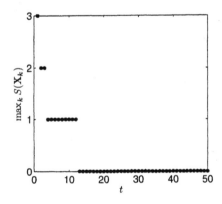

Figure 18.6   Typical evolution of the CE method on the eight-queens problem.

16 (a) Since the permutation flow shop problem is treated as a TSP problem, any CE implementation used to solve any of the problems 8.10, 8.11, or 8.12 can be used to solve the permutation flow shop problem. The objective function is coded in the following m-file:

```
function completion_time=PFSP(x,t)
% inputs: x=(x1,...,xn) is a permutation of jobs;
% t is n by m  matrix where t(i,j) is the processing time
% of job i on machine j;
% output: completion_time=C(x(n),m);
[n,m]=size(t);
C=nan(n,m);% preallocate memory
C(x(1),1)=t(x(1),1); % time to process the x(1)-th job on machine 1
for i=2:n % time taken by machine 1 to process the n jobs
    C(x(i),1)=C(x(i-1),1)+t(x(i),1);
end

for j=2:m % time taken by the  m-th machine to finish job x(1)
    C(x(1),j)=C(x(1),j-1)+t(x(1),j);
end

for i=2:n
    for j=2:m
        C(x(i),j)=max( C(x(i-1),j), C(x(i),j-1) ) + t(x(i),j);
    end
end
completion_time=C(x(n),m);
```

(b) Benchmark problems can, for example, be found on the following website:

http://ina2.eivd.ch/Collaborateurs/etd/problemes.dir/ordonnancement.dir/ordonnancement.html.

For example, consider problem ta002. The best known solution is $x^* = (6, 10, 17, 7, 18, 15, 12, 3, 1, 9, 11, 8, 19, 4, 13, 16, 14, 5, 20, 2)$ with $C(x_{20}^*, 5) = 1359$. The typical evolution of the CE algorithm for this problem is given in Table 18.7. The CE parameters are $N = 10^4$, $\varrho = 10^{-3}$, and $\alpha = 0.7$. The algorithm was stopped when there was no improvement in $\widehat{\gamma}_t$ over three consecutive iterations. Figure 18.7 shows the final transition matrix.

Table 18.7 Evolution of the CE algorithm for the permutation flow shop problem ta002.

t	S_t^*	$\widehat{\gamma}_t$
1	1373	1382
2	1370	1373
3	1367	1373
4	1367	1368
5	1367	1367
6	1367	1367
7	1367	1367

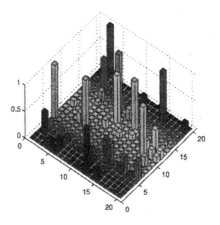

Figure 18.7 Plot of the final transition matrix $\mathbf{P}_{15}$.

8.17   The stochastic CE program which has to be solved at iteration t is:

$$\max_{\mu_t, \sigma_t} \frac{1}{N} \sum_{k=1}^{N} I_{\{S(\mathbf{X}_k) \geqslant \widehat{\gamma}_t\}} \ln f(\mathbf{X}_k; \mu_t, \sigma_t), \quad \{\mathbf{X}_k\} \sim_{iid} \mathsf{N}(\widehat{\mu}_{t-1}, \widehat{\Sigma}_{t-1}),$$

where f is multivariate normal with mean μ_t and diagonal covariance matrix with main diagonal given by $\{\sigma_{ti}^2\}$. If $\mathcal{E}_t$ denotes the collection of $\mathbf{X}_k$ for which $\{S(\mathbf{X}_k) \geqslant \widehat{\gamma}_t\}$ — the set of elite samples —, then the above optimization problem is equivalent to

$$\max_{\mu_t, \sigma_t} \sum_{\mathbf{X}_k \in \mathcal{E}_t} \ln f(\mathbf{X}_k; \mu_t, \sigma_t).$$

Therefore, the updating formulas are simply the maximum likelihood estimators of the mean and the (diagonal) covariance matrix of the normal distribution, based on the elite sample $\mathcal{E}_t = \{\mathbf{X}_k : S(\mathbf{X}_k) \geqslant \widehat{\gamma}_t\}$.

8.18   Typing the command peaks in the Matlab command window produces a plot of the peaks function which shows three local maxima and three local minima.

Peaks function

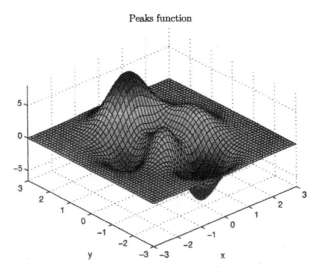

Figure 18.8 The peaks function has 3 local maxima

3.19   A sequence of sampling densities is plotted along with the objective function in Figure 18.9.

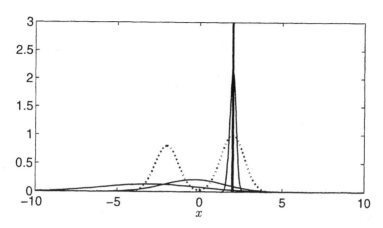

Figure 18.9   Evolution of the sampling densities (solid lines) with objective function (dotted line).

3.20   The following script can be used to minimize the trigonometric function.

```
clear all,clc
N=1000;    % sample size
n=10;      % dimensionality of problem
N_el=20; % elite sample size

sig=100*ones(1,n);%inital scale of Gaussian
mu=100*ones(1,n);
alpha=.8; % smoothing parameter
table=[];
for t=1:4000
    % generate sample
    X=repmat(mu,N,1)+randn(N,n).*repmat(sig,N,1);
       Scores=trigonometric(X);
       gam=prctile(Scores,N_el/N*100);
       Index=(Scores<=gam);


% update mean and covariance based on elite sample
mu=alpha*mean(X(Index,:),1)+(1-alpha)*mu;
sig=alpha*sqrt(var(X(Index,:),1))+(1-alpha)*sig;

if mod(t,5)==0 % record results every 5 iterations
   table=[table; t,gam, mu];
end
  % stopping criterion of the method
 if norm(sig)<10^-6, break, end
end
% display results
table
```

The script uses the following function file.

```
function out=trigonometric(X)
eta=7; mu=1;
out=8*sin(eta*(X-.9).^2).^2+6*sin(2*eta*(X-.9).^2).^2+mu*(X-.9).^2;
out=1+sum(out,2);
```

The graph of the function for $n = 2$ is shown in Figure 18.10. A typical evolution of the algorithm is presented in Table 18.8.

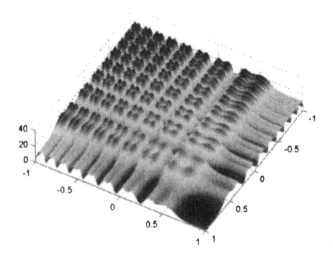

Figure 18.10   Plot of the trigonometric function for $n = 2$.

Table 18.8   The evolution of the CE algorithm with $n = 10$, $N = 1000$, $\alpha = 0.8$, and $N_e = 20$.

t	$\widehat{\gamma}_t$	$\widehat{\mu}_{1,t}$	$\widehat{\mu}_{2,t}$	$\widehat{\mu}_{3,t}$	$\widehat{\mu}_{4,t}$	$\widehat{\mu}_{5,t}$	$\widehat{\mu}_{6,t}$	$\widehat{\mu}_{7,t}$	$\widehat{\mu}_{8,t}$	$\widehat{\mu}_{9,t}$	$\widehat{\mu}_{10,t}$
5	1158.1	0.46955	5.9725	2.0675	6.1391	4.2022	7.5433	3.8765	3.6651	9.888	1.9784
10	62.34	1.4202	1.1508	0.9294	0.81989	0.90251	0.90956	0.73656	0.78576	2.1126	1.1873
15	26.083	0.98987	1.0183	0.94035	0.93537	0.91474	0.90367	0.89705	0.93073	0.96638	0.86246
20	1.0375	0.89361	0.89987	0.89848	0.90016	0.89389	0.89681	0.91033	0.894	0.90227	0.89586
25	1	0.89988	0.8993	0.89934	0.89992	0.90018	0.90069	0.90043	0.90033	0.89981	0.90029
30	1	0.9	0.90001	0.9	0.89994	0.89999	0.89996	0.89999	0.90003	0.89999	0.9
35	1	0.9	0.9	0.9	0.9	0.9	0.9	0.9	0.89999	0.9	0.9
40	1	0.9	0.9	0.9	0.9	0.9	0.9	0.9	0.9	0.9	0.9

8.21   Table 18.9 shows the performance of the CE algorithm applied to the Rosenbrock problem for dimension $n = 10$. In the table, X_t^* is the best overall candidate solution obtained up to iteration t and $S_t^* = S(X_t^*)$. Overall, the speed and accuracy of the algorithm is reduced with an increased n. The success of the algorithm depends on the appropriate size of the noise in the injection heuristic. The following script was used.

```
clear all,clc,format short g
N=1000; % sample size
n=10; % dimensionality of problem
N_el=100; % elite sample size
sig=10*ones(1,n);%inital scale of Gaussian
mu=ones(1,n);
alpha=.1; % smoothing parameter
table=[];
counter=0;
S_best=inf;
for t=1:4000
    X=repmat(mu,N,1)+randn(N,n).*repmat(sig,N,1); %generate sample
    Scores=Rosenbrock(X);
    gam=prctile(Scores,N_el/N*100);
    % update mean and covariance based on elite sample
    Index=(Scores<=gam);
    mu=mean(X(Index,:),1);
    sig=alpha*sqrt(var(X(Index,:),1))+(1-alpha)*sig;
    % stopping criterion of the method
    if sig<10^-3,  sig=100*sig; counter=counter+1;  end
    if S_best>min(Scores);[S_best, I]=min(Scores);X_best=X(I,:); end
    % apply injection a sufficient number of times
    if counter==20 ,break ,end
    if mod(t,500)==0 % record results every 5 iterations
        table=[table; t, S_best, X_best];
    end
end
table % display results
```

The Rosenbrock function is implemented in the file Rosenbrock.m:

```
function out=Rosenbrock(X)
out=sum(100*(X(:,2:end)-X(:,1:end-1).^2).^2+(X(:,1:end-1)-1).^2,2);
```

Table 18.9 The evolution of CE algorithm with injection for $n = 10$, $N = 1,000$, $\alpha = 0.1$, and $N_e = 100$.

t	S_t^*	$x_{1,t}^*$	$x_{2,t}^*$	$x_{3,t}^*$	$x_{4,t}^*$	$x_{5,t}^*$	$x_{6,t}^*$	$x_{7,t}^*$	$x_{8,t}^*$	$x_{9,t}^*$	$x_{10,t}^*$
500	4.1658	0.94393	0.88991	0.79153	0.62752	0.4035	0.17419	0.039253	0.011242	0.010696	-7.4354e-005
1000	2.6735	0.97661	0.9596	0.91919	0.84318	0.72468	0.53198	0.29858	0.095198	0.027125	-0.00052926
1500	0.38294	0.99784	0.99449	0.98907	0.97836	0.95645	0.91493	0.83719	0.70102	0.49334	0.24295
2000	0.038273	0.99954	0.99871	0.99705	0.9946	0.98855	0.97746	0.95575	0.9128	0.83305	0.69381
2500	0.0049379	0.99911	0.99937	0.99935	0.99874	0.99645	0.99267	0.98494	0.97055	0.94184	0.88622
3000	0.00017667	0.99989	0.99982	0.99977	0.99956	0.99921	0.99857	0.99716	0.99436	0.989	0.97801

.22   Using the implementation from Problem 8.21 we obtained the results in the Table 18.10. Injection was essential for achieving accuracy in problem (b). Note that the CPU time should only be used for relative performance assessment across the problems.

Table 18.10   Results for constrained Rosenbrock problems.

Constraints	S_T^*	secs
$\sum_i x_i \leqslant -8$	1517.8	1.9
$\sum_i x_i \geqslant 15$	1.3101	15
$\sum_i x_i \leqslant -8, \ \sum_i x_i^2 \geqslant 15$	1763.8	4.6
$\sum_i x_i \geqslant 15, \ \sum_i x_i^2 \leqslant 22.5$	493.09	8.3

The constraints for (c), for example, are implemented in the following m-file.

```
function out=Rosenbrock(X)
penalty= 10^3*(max(sum(X,2)+8,0) +max(15-sum(X.^2,2),0));
out=sum(100*(X(:,2:end)-X(:,1:end-1).^2).^2+...
                (X(:,1:end-1)-1).^2,2)+penalty;
```

8.23   For a given x_1, the equality constraints require us to solve a nonlinear system of two
equations in two unknowns (x_2 and x_3). There are two pairs of values for x_2 and x_3 that
solve the system, namely

$$(x_2, x_3) = \left(\frac{224 - 32\,x_1 + 2\,r(x_1)}{70}, \ \frac{2\,(28 - 4\,x_1 - r(x_1))}{35} \right)$$

and

$$(x_2, x_3) = \left(\frac{224 - 32\,x_1 - 2\,r(x_1)}{70}, \ \frac{2\,(28 - 4\,x_1 + r(x_1))}{35} \right),$$

where

$$r(x_1) = \sqrt{2989 + 896\,x_1 - 309\,x_1^2}\,.$$

The script below was used to produce the optimal solution

$$\mathbf{x}^* = (3.51212125617781, 0.216987948327608, 3.55217123914157),$$

with $S(\mathbf{x}^*) = 961.715172130052.$

```
clear all,clc,format long g
N=1000;    % sample size
n=3;       % dimensionality of problem
N_el=100;  % elite sample size
sig=10;%inital scale of Gaussian for X1
mu=rand*5; %inital mean of Gaussian for X1
alpha=.1;  % smoothing parameter
table=[];   counter=0;
for t=1:4000
    % generate X1 as per truncated normal
    X1=normtrnd(repmat(mu,N,1),repmat(sig,N,1),0,5);
    % generate X2 and X3 conditional on X1
    % randomly choose one pair of solutions
```

```
    Sign=sign(2*rand(N,1)-1);
    X2=(224-32*X1+2*Sign.*sqrt(2989+896*X1-309*X1.^2))/70;
    X3=2/35*(28-4*X1-Sign.*sqrt(2989+896*X1-309*X1.^2));
    X=[X1,X2,X3];
    Scores=S(X);
    gam=prctile(Scores,N_el/N*100);
    Index=(Scores<=gam);
    [S_best,I]=min(Scores);
    X_best=X(I,:);
    % update mean and covariance based on elite sample
    mu=alpha*mean(X1(Index))+(1-alpha)*mu;
    sig=alpha*sqrt(var(X1(Index)))+(1-alpha)*sig;

    % stopping criterion of the method
    if sig(1)<10^-6,  sig=100*sig; counter=counter+1;  end
    % apply injection a sufficient number of times
    if counter==1 ,break ,end
    if mod(t,5)==0 % record results every 5 iterations
        table=[table; t, S_best, X_best];
    end
end
% display results
table
```

The script uses a function that generates random variables from the truncated normal distribution using the inverse-transform method and a function which encodes the score.

```
 function result = normtrnd(mu,std,left,right)
% PURPOSE: random draws from a  truncated normal
% ----------------------------------------------------
% USAGE:    y = normtrnd(mu,sigma2,left,right)
% where:  mu = mean (nobs x 1)
%          std = standard deviation (nobs x 1)
%         left = left truncation points (nobs x 1)
%        right = right truncation points (nobs x 1)
% ----------------------------------------------------
% RETURNS: y = (nobs x 1) vector
% ----------------------------------------------------

% Calculate bounds on probabilities
  lowerProb = Phi((left-mu)./std);
  upperProb = Phi((right-mu)./std);
% Draw uniform from within (lowerProb,upperProb)
  u = lowerProb+(upperProb-lowerProb).*rand(size(mu));
% Find needed quantiles
  result = mu + Phiinv(u).*std;
function val=Phiinv(x)
% Computes the standard normal quantile function of the vector x.
```

```
val=sqrt(2)*erfinv(2*x-1);
function y = Phi(x)
% Phi computes the standard normal distribution function value at x
y = .5*(1+erf(x/sqrt(2)));
```

```
function out=S(x)
penalty=10^6*max(-x(:,2),0)+10^6*max(-x(:,3),0) ;
out=10^3-x(:,1).^2-2*x(:,2).^2-x(:,3).^2-...
            x(:,1).*x(:,2)-x(:,1).*x(:,3)+penalty;
```

8.24   Figure 18.11 shows the behavior of $\hat{\gamma}_t$, $\hat{\mu}_t$, and $\hat{\sigma}_t$ for the cases with U($-0.1, 0.1$) (left) and N(0, 0.01) (right) noise, respectively. Figure 18.12 shows the behavior for the case with the larger N(0, 1) noise. In all cases the convergence criterion was to stop when $\hat{\sigma}_t \leqslant 10^{-5}$.

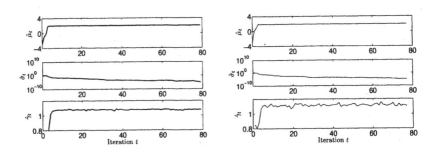

Figure 18.11   Convergence behavior with U($-0.1, 0.1$) (left) and N(0, 0.01) (right) noise.

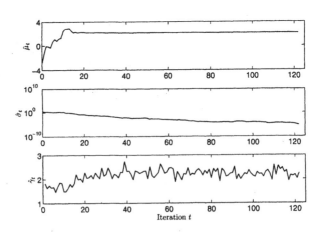

Figure 18.12   Convergence behavior with (large) N(0, 1) noise.

8.25   Figure 18.13 depicts a sequence of mean vectors from a run of the CE algorithm on Matlab's peaks function, when noise is added. Figure 18.14 shows the evolution of the best and worst scores in the elite sample, as well as the scores of the sequence of mean vectors.

Finally, Figure 18.15 shows the Euclidean distance between the mean vector and the global optimum from iteration to iteration.

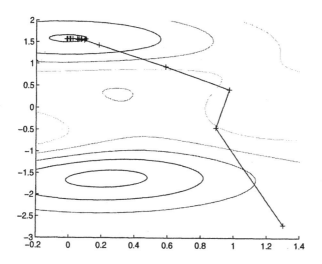

Figure 18.13 Evolution of the mean vector for the noisy peaks function.

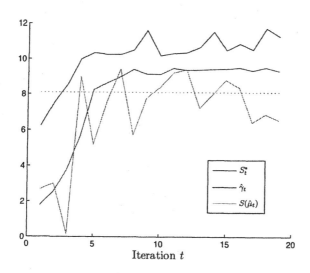

Figure 18.14 Best and worst of the elite scores, along with the score of the mean vector.

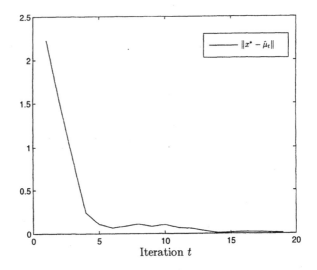

Figure 18.15   The noisy mean vector converges to the true optimal solution.

8.26   We chose the following parameters for this (noisy) synthetic TSP: $a = 1.5$, $b = 3$ and $n = 30$. For the CE parameters we chose $N = 4500$, $N_e = 135$, and $\alpha = 0.7$. We ran the (noisy) CE algorithm ten times. Nine of the ten trials found the true solution. Figure 18.16 illustrates the evolution of the algorithm for the nine runs that found the true solution. Compare this with Figure 18.17, where the same results are displayed for the deterministic counterpart. With exactly the same parameters, but without noise, the CE algorithm found the true solution in all ten trials, converging in about 12 iterations, as opposed to 60 for the noisy case. The rate of convergence also appears to be more consistent.

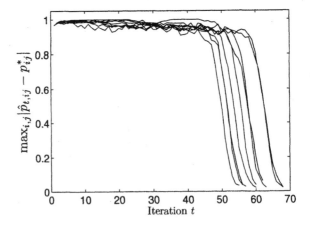

Figure 18.16   The evolution of nine runs of the CE algorithm on a noisy TSP instance.

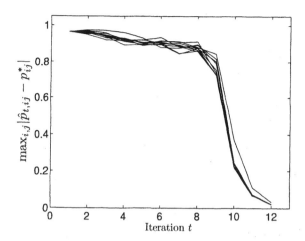

Figure 18.17 The evolution of ten runs of the CE algorithm on a TSP instance without noise.

CHAPTER 19

COUNTING VIA MONTE CARLO

9.1   Let $Y = X_1 + \cdots + X_n$, where X_i are iid $\mathrm{Ber}(p)$ random variables, and let $\ell_L = \mathbb{P}(Y \leqslant (1-\varepsilon)np)$. Recall the Chernoff bound **(9.18)**:

$$\mathbb{P}(Y \leqslant a) \leqslant \min_{\theta > 0} \left\{ e^{\theta a}\, \mathbb{E}[e^{-\theta Y}] \right\} .$$

In this case, we have

$$\mathbb{P}(Y \leqslant (1-\varepsilon)np) \leqslant \min_{\theta > 0} \left\{ e^{\theta(1-\varepsilon)np}\, \mathbb{E}[e^{-\theta Y}] \right\}$$
$$= \min_{\theta > 0} \left\{ e^{\theta(1-\varepsilon)np}\, \left(\mathbb{E}[e^{-\theta X}] \right)^n \right\} ,$$

where $X \sim \mathrm{Ber}(p)$. Now, $\mathbb{E}[e^{-\theta X}] = (1-p) + pe^{-\theta}$, so that

$$\mathbb{P}(Y \leqslant (1-\varepsilon)np) \leqslant \min_{\theta > 0} \left\{ e^{\theta(1-\varepsilon)\,np} \left((1-p) + pe^{-\theta} \right)^n \right\} .$$

Taking logarithms of both sides, we see that

$$\ln(\ell_L) \leqslant \min_{\theta > 0} \left\{ \theta\,(1-\varepsilon)\,np + n\ln\left((1-p) + pe^{-\theta} \right) \right\} .$$

To solve the minimization problem on the right-hand side, take the derivative of the upper bound with respect to θ and set it to zero. This gives

$$(1-\varepsilon)\,np - \frac{np\,e^{-\theta}}{1-p+np\,e^{-\theta}} = 0 .$$

The solution to this equation is

$$\theta^* = \ln\left(\frac{1-p+p\varepsilon}{(1-p)(1-\varepsilon)}\right),$$

which is positive, provided that $\varepsilon > 0$. Substituting θ^* into the upper bound, we find

$$\ln(\ell_L) \leqslant \theta^*(1-\varepsilon)np + n\ln\left((1-p) + p\frac{(1-p)(1-\varepsilon)}{1-p+p\varepsilon}\right)$$

$$= \theta^*(1-\varepsilon)np + n\ln\left(\frac{(1-p)(1-p+p\varepsilon) + p(1-p)(1-\varepsilon)}{1-p+p\varepsilon}\right)$$

$$= \theta^*(1-\varepsilon)np - n\ln\left(\frac{1-p+p\varepsilon}{(1-p)(1-p+p\varepsilon) + p(1-p)(1-\varepsilon)}\right)$$

$$= \theta^*(1-\varepsilon)np - n\ln\left(\frac{1-p+p\varepsilon}{(1-p)}\right)$$

$$= \theta^*(1-\varepsilon)np - n\ln\left(1 + \frac{p\varepsilon}{(1-p)}\right)$$

$$= np\,h(\varepsilon, p),$$

where $h(\varepsilon, p)$ is given by

$$h(\varepsilon, p) = -\frac{1}{p}\ln\left(1 + \frac{\varepsilon p}{1-p}\right) + (1-\varepsilon)\theta^*.$$

Therefore, we have demonstrated the required upper bound.

9.2   Let $\ell_U = \mathbb{P}(Y \geqslant (1+\varepsilon)np) = \mathbb{P}(-Y \leqslant -(1+\varepsilon)np)$, where Y is as in Problem 9.1. Applying Chernoff's bound yields

$$\mathbb{P}(-Y \leqslant -(1+\varepsilon)np) \leqslant \min_{\theta>0} e^{-\theta(1+\varepsilon)np}\,\mathbb{E}[e^{\theta Y}]$$

$$= \min_{\theta>0} e^{-\theta(1+\varepsilon)np}\left(\mathbb{E}[e^{\theta X}]\right)^n,$$

where $X \sim \mathrm{Ber}(p)$. Since $\mathbb{E}[e^{\theta X}] = (1-p) + pe^{\theta}$, we have

$$\ell_U \leqslant \min_{\theta>0}\left\{e^{-\theta(1+\varepsilon)np}\left((1-p) + pe^{\theta}\right)^n\right\}.$$

Taking logarithms (as in the solution of Problem 9.1) gives

$$\ln(\ell_U) \leqslant \min_{\theta>0}\left\{-\theta(1+\varepsilon)np + n\ln\left((1-p) + pe^{\theta}\right)\right\}.$$

Taking the derivative of the upper bound with respect to θ and setting it to 0 yields

$$-(1+\varepsilon)np + \frac{npe^{\theta}}{1-p+npe^{\theta}} = 0.$$

This has solution

$$\tilde{\theta} = \ln\left(\frac{(1-p)(1+\varepsilon)}{1-p(1+\varepsilon)}\right),$$

which is positive as long as $\varepsilon > 0$. Plugging this back into the upper bound gives

$$\ln(\ell_U) \leqslant \min_{\theta > 0} -\widetilde{\theta}(1 + \varepsilon)\, np + n\ln\left((1 - p) + p\,\frac{(1 - p)(1 + \varepsilon)}{1 - p(1 + \varepsilon)}\right)$$

$$= -\widetilde{\theta}(1 + \varepsilon)\, np + n\ln\left(\frac{1 - p}{1 - p(1 + \varepsilon)}\right)$$

$$= -\widetilde{\theta}(1 + \varepsilon)\, np - n\ln\left(1 - \frac{p\varepsilon}{1 - p}\right)$$

$$= np\widetilde{h}(\varepsilon, p),$$

where

$$\widetilde{h}(\varepsilon, p) = -\frac{1}{p}\ln\left(1 - \frac{p\varepsilon}{1 - p}\right) - (1 + \varepsilon)\widetilde{\theta}.$$

For fixed $\varepsilon \in (0, 1)$,

$$\frac{\partial}{\partial p}\,\widetilde{h}(\varepsilon, p) = \frac{1}{p^2}\left[\ln\left(1 - \frac{\varepsilon p}{1 - p}\right) + \frac{\varepsilon p}{1 - p}\right] < 0,$$

since $y + \ln(1 - y) < 0$ for $y > 0$. We conclude that $\widetilde{h}(\varepsilon, p)$ is monotonically decreasing in p for $p \in (0, 1)$ and fixed $\varepsilon \in (0, 1)$. Therefore,

$$\widetilde{h}(\varepsilon, p) \leqslant \widetilde{h}(\varepsilon, 0) = \varepsilon - (1 + \varepsilon)\ln(1 + \varepsilon)$$

for $p \in (0, 1)$ and fixed $\varepsilon \in (0, 1)$.

A Padé approximant of order (m, n) of a function $f(x)$ is a rational function

$$g(x) = \frac{\sum_{i=0}^{m} a_i x^i}{1 + \sum_{j=1}^{n} b_j x^j}.$$

The coefficients in g are chosen so that the first $m + n + 1$ coefficients of the Taylor expansion of g agree with the corresponding ones of f. We find the Padé approximant of order $(2, 1)$ of $f(\varepsilon) = \varepsilon - (1 + \varepsilon)\ln(1 + \varepsilon)$ by equating the first four coefficients of the Taylor expansion of $f(\varepsilon)$ around $\varepsilon = 0$ with the first four coefficients of the Taylor expansion of $g(\varepsilon)$ around $\varepsilon = 0$, where $g(\varepsilon)$ is given by

$$\frac{a + b\varepsilon + c\varepsilon^2}{1 + d\varepsilon}.$$

This process yields $a = b = 0$, $c = -1/2$, and $d = 1/3$. Plugging these in, we have

$$\varepsilon - (1 + \varepsilon)\ln(1 + \varepsilon) \approx -\frac{\frac{\varepsilon^2}{2}}{1 + \frac{\varepsilon}{3}} = -\frac{\varepsilon^2}{2 + \frac{2\varepsilon}{3}}.$$

We can verify that indeed $f(\varepsilon) \leqslant g(\varepsilon)$ for $\varepsilon \in (0, 1)$ as follows. The inequality

$$\varepsilon - (1 + \varepsilon)\ln(1 + \varepsilon) \leqslant -\frac{\varepsilon^2}{2 + \frac{2\varepsilon}{3}}$$

implies that

$$\left(2 + \frac{2\varepsilon}{3}\right)\varepsilon - \left(2 + \frac{2\varepsilon}{3}\right)(1 + \varepsilon)\ln(1 + \varepsilon) \leqslant -\varepsilon^2,$$

which is equivalent to

$$6\varepsilon + \varepsilon^2 - \left(6 + 8\varepsilon + 2\varepsilon^2\right)\ln(1+\varepsilon) \leqslant 0 \, .$$

The derivative of the left-hand side is

$$\frac{\partial}{\partial\varepsilon}\left[6\varepsilon + \varepsilon^2 - \left(6 + 8\varepsilon + 2\varepsilon^2\right)\ln(1+\varepsilon)\right] = -4(2+\varepsilon)\ln(1+\varepsilon) < 0 \quad \text{if } \varepsilon > 0.$$

Hence,

$$6\varepsilon + \varepsilon^2 - \left(6 + 8\varepsilon + 2\varepsilon^2\right)\ln(1+\varepsilon)$$

is monotonically decreasing in ε and the value at $\varepsilon = 0$ is 0. Thus, $f(\varepsilon) \leqslant g(\varepsilon)$ for $\varepsilon \in (0,1)$. Therefore,

$$\ln(\ell_U) \leqslant n\widetilde{ph}(\varepsilon,p) \leqslant n\widetilde{ph}(\varepsilon,0) \leqslant -\frac{np\,\varepsilon^2}{2+\frac{2\varepsilon}{3}} \, ,$$

and so

$$\ell_U \leqslant \exp\left(-\frac{np\,\varepsilon^2}{2+\frac{2}{3}\varepsilon}\right)$$

for all $\varepsilon \in (0,1)$ and $p \in (0,1)$.

9.3   Consider the synthetic max-cut instance in Problem 8.9 with $n = 400$ nodes. The following Matlab script provides an MinxEnt algorithm to solve this max-cut problem.

```
clear all, format short g
% first create the synthetic max-cut cost matrix
m=200; n=400; a=0; b=1; c=1; % set up dimension of C and paramters
Z_1=tril(a+b*rand(m,m),-1);Z_1=Z_1+Z_1';
Z_2=tril(a+b*rand(n-m,n-m),-1);Z_2=Z_2'+Z_2;
C=[Z_1,c*ones(m,n-m);c*ones(n-m,m),Z_2]; % cost matrix
p_star=[ones(1,m),zeros(1,n-m)]; % this is the optimal  cut vector
gamma_star=c*m*(n-m); %the optimal value of the cut
rho=0.1; % rarity parameter
N=1000; % sample size
p=[1,0.5*ones(1,n-1)]; % initial cutting probabilities
tab=[];
lam=[0;0];
for iter=1:400 % set up iteration counter
  X=(rand(N,n)<repmat(p,N,1)); % create N number of cut vectors
  [Scores,Index]=sort(S(X,C),'descend'); X=X(Index,:);
  gam=mean(Scores(1:floor(rho*N)));
  % determine the lagrange multipliers of the MinxEnt  program
  convex_equation=@(lam)mce_pdf(lam,Scores,gam);
  lam = fsolve(convex_equation,lam,...
          optimset('Display','off','Jacobian','on'));
  mce=exp(lam(1)+lam(2)*Scores);
  % update the probabilities based on the MinxEnt formula
  p=mce'*X; p(1)=1;
  % difference between optimal solution and CE solution
```

```
    error=norm(p-p_star);
    tab=[tab;iter,gam,max(Scores),error];
    if abs(error)<10^-3, break , end
end
tab
```

The script uses the S.m function from Problem 8.9 and the following function m-file required for the solution of the MinxEnt program.

```
function [fun, jacobian]=mce_pdf(lambdas,Scores,gamma,weights)
%computing the unique root of the two-dimensional function mce_pdf
% yields the Lagrange multipliers of the MinxEnt program
% for efficiency, mce_pdf  provides the exact Jacobian matrix
% to the Matlab root-finding routine fsolve.m
if nargin<4
    weights=zeros(size(Scores));
end
A=sum(exp(lambdas(1)+lambdas(2)*Scores+weights));
B=sum(exp(lambdas(1)+lambdas(2)*Scores+weights).*Scores);
fun=[1-A;
     gamma-B ];
 if nargout>1
    jacobian=[-A, -B;...
             -B, -sum(exp(lambdas(1)+...
                    lambdas(2)*Scores+weights).*Scores.^2) ];
end
```

In this case, the convergence of the PME algorithm is slower than that of the CE algorithm.

Table 19.1 The evolution of the PME algorithm for the max-cut problem with $n = 400$, $m = 200$, $\varrho = 0.1$, $N = 1000$. Note that by convention $x_1 \in V_1$ so that $\widehat{p}_{1,t} = 1$ for all t.

t	$\widehat{\gamma}_t$	$\max_k S(\mathbf{X}_{k,t})$	$\|\widehat{\mathbf{p}}_t - \mathbf{p}\|$
1	30105	30264	10.002
5	30108	30343	10.013
10	30103	30342	9.9651
15	30168	30363	9.4916
20	31216	31700	6.9145
25	34948	35676	3.6783
30	38789	39138	1.0909
35	40000	40000	0.070575
40	40000	40000	0.0050437
45	40000	40000	0.0004945

).4 Consider an arbitrary collection of n subsets $A_1, A_2, \ldots, A_n$ of some finite set $\mathscr{X}$. Let B_k be set of all distinct k-fold intersections of the $\{A_i\}$ for $k = 1, 2, \ldots n$. Then, we wish to prove

$$\left| \bigcup_{i=1}^{n} A_i \right| = \sum_{k=1}^{n} (-1)^{k+1} \sum_{C \in B_k} |C|.$$

For the case $n = 1$, we have trivially $|A_1| = |A_1|$. For $n = 2$, we note that $A_1 \cup A_2$ can be written as $(A_1 \cap A_2^c) \cup (A_1^c \cap A_2) \cup (A_1 \cap A_2)$, where each term in the union is disjoint. Hence,

$$|A_1 \cup A_2| = |A_1 \cap A_2^c| + |A_1^c \cap A_2| + |A_1 \cap A_2|.$$

Further, notice that $|A_1| = |A_1 \cap A_2^c| + |A_1 \cap A_2|$ and that $|A_2| = |A_1 \cap A_2^c| + |A_1 \cap A_2|$, since $A_1 = (A_1 \cap A_2^c) \cup (A_1 \cap A_2)$ and $A_2 = (A_1 \cap A_2^c) \cup (A_1 \cap A_2)$, where the unions are of disjoint sets. Hence,

$$|A_1 \cap A_2^c| + |A_1^c \cap A_2| + |A_1 \cap A_2| = |A_1| - |A_1 \cap A_2| + |A_2| - |A_1 \cap A_2| + |A_1 \cap A_2|$$
$$= |A_1| + |A_2| - |A_1 \cap A_2|.$$

This is precisely the inclusion–exclusion principle for $n = 2$.

Next, assume that the inclusion–exclusion principle holds for $n = m \geqslant 2$. We shall show that it holds for $n = m + 1$ as well. We have

$$\left| \bigcup_{i=1}^{m+1} A_i \right| = \left| \left(\bigcup_{i=1}^{m} A_i \right) \cup A_{m+1} \right| = \left| \bigcup_{i=1}^{m} A_i \right| + |A_{m+1}| - \left| \left(\bigcup_{i=1}^{m} A_i \right) \cap A_{m+1} \right|,$$

where we have used the inclusion–exclusion principle for $n = 2$. Note that

$$\left(\bigcup_{i=1}^{m} A_i \right) \cap A_{m+1} = \bigcup_{i=1}^{m} (A_i \cap A_{m+1}),$$

so that by applying the case $n = m$ — which is assumed to be true —, we have

$$\left| \left(\bigcup_{i=1}^{m} A_i \right) \cap A_{m+1} \right| = \sum_{k=1}^{m} (-1)^{k+1} \sum_{C \in B_k} |C \cap A_{m+1}|,$$

where B_k is the set of all distinct k-fold intersections of $A_1, \ldots, A_m$.

Hence, we have

$$\left| \bigcup_{i=1}^{m+1} A_i \right| = \sum_{k=1}^{m} (-1)^{k+1} \sum_{C \in B_k} |C| + |A_{m+1}| - \sum_{k=1}^{m} (-1)^{k+1} \sum_{C \in B_k} |C \cap A_{m+1}|,$$

where B_k is the set of all distinct k-fold intersections of $A_1, \ldots, A_m$. Notice that

$$\sum_{C \in B_k} |C \cap A_{m+1}| = \sum_{C \in D_{k+1}} |C|,$$

where D_k is the set of all distinct k-fold intersections of $A_1, \ldots, A_{m+1}$ that contain A_{m+1}. Further, let E_k be the set of all distinct k-fold intersections of $A_1, \ldots, A_{m+1}$. Then, $E_k = B_k \cup D_k$, with $B_k \cap D_k = \emptyset$. Finally, observe that $D_1 = A_{m+1}$ and $B_{m+1} = \emptyset$.

Using this knowledge, we have

$$\left| \bigcup_{i=1}^{m+1} A_i \right| = |A_{m+1}| + \sum_{k=1}^{m}(-1)^{k+1} \sum_{C \in B_k} |C| - \sum_{k=1}^{m}(-1)^{k+1} \sum_{C \in D_{k+1}} |C|$$

$$= |A_{m+1}| + \sum_{k=1}^{m}(-1)^{k+1} \sum_{C \in B_k} |C| + \sum_{k=2}^{m+1}(-1)^{k+1} \sum_{C \in D_k} |C|$$

$$= \sum_{k=1}^{m+1}(-1)^{k+1} \left(\sum_{C \in B_k} |C| + \sum_{C \in D_k} |C| \right)$$

$$= \sum_{k=1}^{m+1}(-1)^{k+1} \sum_{C \in E_k} |C| \; .$$

Thus, if the inclusion–exclusion principle holds for $n = m$ and $n = 2$, then it holds for $n = m + 1$. As we have shown the case $n = 2$ holds, we have by mathematical induction that the inclusion–exclusion principle holds for any $n \geqslant 2$. Since we have shown it holds trivially when $n = 1$, we have that the inclusion–exclusion principle holds for all $n \geqslant 1$.

.5 (a) In this problem we have $\mathscr{A} = \{1, 2, 3, 4, 5\}$, $\mathscr{A}_1 = \{1, 2, 5\}$, $\mathscr{A}_2 = \{1, 4\}$, $\mathscr{A}_3 = \{3, 5\}$, $\mathscr{A}_4 = \{3, 4\}$, and $\mathscr{A}_5 = \{1\}$. We wish to count the total number of distinct representatives $\mathbf{x} = (x_1, \ldots, x_5)$, where $x_k \in \mathscr{A}_k$, $k = 1, 2, \ldots, 5$. Firstly, notice that x_5 must be set to 1. This in turn dictates that x_2 must be set to 4, which forces x_4 to be 3, which sets x_3 to 5, which finally requires that x_1 be set to 2. Hence, for this problem, there is only one way to select distinct members from each set. Therefore, the total number of distinct representatives is 1, and the representative is $\mathbf{x} = (2, 4, 5, 3, 1)$.

(b) The permanent of an $n \times n$ matrix A is given by

$$\text{perm}(A) = \sum_{\mathbf{x} \in \mathscr{X}} \prod_{i=1}^{n} a_{ix_i} \; ,$$

where $\mathscr{X}$ is the set of all permutations of $\{1, 2, \ldots, n\}$.

For a matrix of 0s and 1s this corresponds to selecting one 1 in each row, such that each column has exactly one 1. For the matrix

$$A = \begin{pmatrix} 1 & 1 & 0 & 0 & 1 \\ 1 & 0 & 0 & 1 & 0 \\ 0 & 0 & 1 & 0 & 1 \\ 0 & 0 & 1 & 1 & 0 \\ 1 & 0 & 0 & 0 & 0 \end{pmatrix} \; ,$$

this corresponds to the number of ways of choosing one element from $\mathscr{A}_1 \in \{1, 2, 5\}$, $\mathscr{A}_2 \in \{1, 4\}$, $\mathscr{A}_3 \in \{3, 5\}$, $\mathscr{A}_4 \in \{3, 4\}$, and $\mathscr{A}_5 \in \{1\}$, in such a way that every element is represented exactly once. Thus, finding the permanent of this matrix of 0s and 1s corresponds directly to finding the total number of distinct representatives in the above problem.

).6 Let $X_1, \ldots, X_n$ be independent random variables, each with marginal pdf f. Suppose we wish to estimate $\ell = \mathbb{P}_f(X_1 + \cdots + X_n \geqslant \gamma)$ using MinxEnt. For the prior pdf one could choose $h(\mathbf{x}) = f(x_1)f(x_2)\cdots f(x_n)$, that is, the joint pdf. We consider only a single constraint in the MinxEnt program, namely $S(\mathbf{x}) = x_1 + \cdots + x_n$. The solution to this program is given by

$$g(\mathbf{x}) = c\,h(\mathbf{x})\,e^{\lambda S(\mathbf{x})} = c\prod_{j=1}^{n} e^{\lambda x_j}\,f(x_j)\,,$$

where $c = 1/\mathbb{E}_h[e^{\lambda S(\mathbf{X})}] = (\mathbb{E}_f[e^{\lambda X}])^{-n}$ is a normalization constant and λ satisfies

$$\frac{\mathbb{E}_h[S(\mathbf{X})e^{\lambda S(\mathbf{X})}]}{\mathbb{E}_h[e^{\lambda S(\mathbf{X})}]} = \gamma\,.$$

The k-th marginal distribution is given by

$$g(x_k) = \int_{\mathbf{x}_{-k}} (\mathbb{E}_f[e^{\lambda X}])^{-n} \prod_{j=1}^{n} e^{\lambda x_j}\,f(x_j)\,\mathrm{d}\mathbf{x}_{-k}\,,$$

where $\mathbf{x}_{-k} = (x_1, \ldots, x_{k-1}, x_{k+1}, \ldots, x_n)$. This becomes

$$g(x_k) = (\mathbb{E}_f[e^{\lambda X}])^{-1}\,e^{\lambda x_k}\,f(x_k)\,,$$

since $g(\mathbf{x})$ is seen to be the product of its n marginal pdfs. Finally, we observe that $g(x_k)$ is the old marginal pdf $f(x_k)$ exponentially tilted with tilting parameter λ.

9.7   Suppose that $S(\mathbf{x})$ is a coordinatewise separable function, that is,

$$S(\mathbf{x}) = \sum_{k=1}^{n} S_k(x_k)\,.$$

Suppose further that $\{X_i\}$ are independent under the prior pdf h. In this case, the optimal MinxEnt pdf is given by

$$\begin{aligned}
g(\mathbf{x}) &= \frac{h(\mathbf{x})\,e^{\lambda S(\mathbf{x})}}{\mathbb{E}_h[e^{\lambda S(\mathbf{X})}]}\\
&= \frac{h(\mathbf{x})\prod_{k=1}^{n} e^{\lambda S_k(x_k)}}{\mathbb{E}_h[\prod_{k=1}^{n} e^{\lambda S_k(X_k)}]}\\
&= h(\mathbf{x})\prod_{k=1}^{n} \frac{e^{\lambda S_k(x_k)}}{\mathbb{E}_h[e^{\lambda S_k(X_k)}]}\,,
\end{aligned}$$

where the last equality holds because the $\{X_i\}$ are independent under the prior h. Since the $\{X_i\}$ are independent under h, we may write $h(\mathbf{x}) = h_1(x_1) \cdots h_n(x_n)$, to find

$$g(\mathbf{x}) = \prod_{k=1}^{n} \frac{h_k(x_k)e^{\lambda S_k(x_k)}}{\mathbb{E}_h[e^{\lambda S_k(X_k)}]}\,.$$

Each term in the product is a pdf in its own right and all the terms in the product involve only a single component of $\mathbf{x}$. As such, $g(\mathbf{x})$ is directly seen to be the product of its marginals, with the k-th marginal of $g(\mathbf{x})$ given by

$$g(x_k) = \frac{h_k(x_k)e^{\lambda S_k(x_k)}}{\mathbb{E}_h[e^{\lambda S_k(X_k)}]}\,,\quad k = 1, 2, \ldots, n\,.$$

9.8   Let $\mathscr{X}$ be the set of permutations $\mathbf{x} = (x_1, \ldots, x_n)$ of the numbers $1, \ldots, n$ and let

$$S(\mathbf{x}) = \sum_{j=1}^{n} j\, x_j \,. \tag{19.1}$$

Let $\mathscr{X}^* = \{\mathbf{x} : S(\mathbf{x}) \geq \gamma\}$, where γ is chosen such that $|\mathscr{X}^*|$ is very small relative to $|\mathscr{X}| = n!$. We seek to estimate $|\mathscr{X}^*|$ using the form

$$|\mathscr{X}^*| = |\mathscr{X}_0| \prod_{k=1}^{n} \frac{|\mathscr{X}_k|}{|\mathscr{X}_{k-1}|} \,,$$

where $\mathscr{X}_j = \{\mathbf{x} : S(\mathbf{x}) \geq \gamma_j\}$ for some sequence of $\{\gamma_j\}$ with $0 = \gamma_0 < \gamma_1 < \cdots < \gamma_r = \gamma$. We can use, for example,

$$\widehat{|\mathscr{X}^*|} = |\mathscr{X}_0| \prod_{k=1}^{r} \widehat{\eta}_k \,,$$

where $\widehat{\eta}_k$ estimates

$$\frac{|\mathscr{X}_k|}{|\mathscr{X}_{k-1}|} = \mathbb{P}_{\mathsf{U}}(\mathbf{X} \in \mathscr{X}_k \,|\, \mathbf{X} \in \mathscr{X}_{k-1}) \,.$$

Here U is the uniform distribution on $\mathscr{X}_{k-1}$. We use the Metropolis–Hastings algorithm to sample from the uniform distribution on $\mathscr{X}_{k-1}$, with $\mathbf{x}$ and $\mathbf{y}$ considered neighbors if one can be obtained from the other by swapping two indices.

Suppose we wish to estimate $\mathbb{P}(S(\mathbf{X}) \geqslant 204)$ for $n = 8$. It is known that this probability is exactly $1/8!$, so that $|\mathscr{X}^*| = 1$. Suppose we choose our sequence of γs to be $(0, 51, 102, 153, 204)$ and suppose further that we estimate each proportion $\widehat{\eta}_k$ using $N = 10{,}000$ samples drawn using the Metropolis–Hastings algorithm. Denote the partial products by

$$\widehat{\ell}_k = \prod_{j=1}^{k} \widehat{\eta}_j \,.$$

A single run of our algorithm yielded the following table.

γ	0	51	102	153	204
$\widehat{\eta}_k$	-	1	1	0.7037	0.0001
$\widehat{\ell}_k$	-	1	1	0.7037	7.0370e-5

This gives a final estimate for $\widehat{|\mathscr{X}^*|}$ of $8! \times 7.0370 \cdot 10^{-5} \approx 2.8373$.

9.9   The primal program which has to be solved is

$$\min_{g} \quad \int g(\mathbf{x}) \ln \frac{g(\mathbf{x})}{h(\mathbf{x})}\, d\mathbf{x}$$

$$\text{subject to:} \quad \int S_i(\mathbf{x})\, g(\mathbf{x})\, d\mathbf{x} = \gamma_i, \quad i = 1, \ldots, m\,,$$

$$\int g(\mathbf{x})\, d\mathbf{x} = 1\,,$$

$$-\int S_i(\mathbf{x})\, g(\mathbf{x})\, d\mathbf{x} \leqslant -\gamma_i, \quad i = m+1, \ldots, m+M\,.$$

The Lagrangian here is

$$\mathcal{L}(g, \lambda, \beta) = \int g(\mathbf{x}) \ln \frac{g(\mathbf{x})}{h(\mathbf{x})} d\mathbf{x} + \sum_{i=1}^{m+M} \lambda_i \left(\gamma_i - \int S_i(\mathbf{x}) g(\mathbf{x}) d\mathbf{x} \right) + \beta \left(\int g(\mathbf{x}) d\mathbf{x} - 1 \right)$$

$$= -\beta + \sum_{i=1}^{m+M} \lambda_i \gamma_i + \int \left(g(\mathbf{x}) \ln \frac{g(\mathbf{x})}{h(\mathbf{x})} - \sum_{i=1}^{m+M} \lambda_i S_i(\mathbf{x}) g(\mathbf{x}) + \beta g(\mathbf{x}) \right) d\mathbf{x} ,$$

and the Lagrange dual program is

$$\max_{\lambda, \beta} \quad \inf_{g \in \mathscr{G}} \mathcal{L}(g, \lambda, \beta)$$
$$\text{subject to:} \quad \lambda_i \geqslant 0, \quad i = m + 1, \ldots, M .$$

Here, $\mathscr{G}$ is the set of admissible pdfs. To compute $\inf_{g \in \mathscr{G}} \mathcal{L}(g, \lambda, \beta)$, one has to use the *Euler-Lagrange* equations in the theory of *calculus of variations*. These equations call for the solution of

$$\frac{\partial}{\partial g} \left(g(\mathbf{x}) \ln \frac{g(\mathbf{x})}{h(\mathbf{x})} - \sum_{i=1}^{m+M} \lambda_i S_i(\mathbf{x}) g(\mathbf{x}) + \beta g(\mathbf{x}) \right) = 0$$

for each $\mathbf{x}$. The solution is

$$g(\mathbf{x}) = h(\mathbf{x}) \exp \left(-1 - \beta + \sum_{i=1}^{m+M} \lambda_i S_i(\mathbf{x}) \right) ,$$

which after substitution gives

$$\inf_{g \in \mathscr{G}} \mathcal{L}(g, \lambda, \beta) = -\beta + \sum_{i=1}^{m+M} \lambda_i \gamma_i - \int h(\mathbf{x}) \exp \left(-1 - \beta + \sum_{i=1}^{m+M} \lambda_i S_i(\mathbf{x}) \right) d\mathbf{x} .$$

Hence, the dual can be written as

$$\max_{\lambda, \beta} \quad -\beta + \sum_{i=1}^{m+M} \lambda_i \gamma_i - \int h(\mathbf{x}) \exp \left(-1 - \beta + \sum_{i=1}^{m+M} \lambda_i S_i(\mathbf{x}) \right) d\mathbf{x}$$
$$\text{subject to:} \quad \lambda_i \geqslant 0, \quad i = m + 1, \ldots, M .$$

CHAPTER 20

APPENDIX

1   From the definition of conditional probability (**1.3**) it follows that

$$
\mathbb{P}\left(X_1 = x_1, \ldots, X_n = x_n \;\middle|\; \sum_{i=1}^n X_i = k \right)
$$

$$
\propto \mathbb{P}\left(X_1 = x_1, \ldots, X_n = x_n, \sum_{i=1}^n X_i = k \right)
$$

$$
= \mathbb{P}\left(X_1 = x_1, \ldots, X_n = x_n, \sum_{i=1}^n x_i = k \right)
$$

$$
= I_{\{\sum_{i=1}^n x_i = k\}} \prod_{i=1}^n p_i^{x_i} (1 - p_i)^{1 - x_i}
$$

$$
\propto I_{\{\sum_{i=1}^n x_i = k\}} \prod_{i=1}^n w_i^{x_i} \;.
$$

A.2 (a) Using the fact that $\Sigma_{12} = \Sigma_{21}$ and $\Sigma_{22}^T = \Sigma_{22}$, direct verification shows that

$$\Sigma \begin{pmatrix} I & 0 \\ -S^T & I \end{pmatrix} = \begin{pmatrix} \Sigma_{11} - \Sigma_{12}S^T & \Sigma_{12} \\ \Sigma_{21} - \Sigma_{22}S^T & \Sigma_{22} \end{pmatrix} = \begin{pmatrix} \Sigma_{11} - \Sigma_{12}S^T & \Sigma_{12} \\ 0 & \Sigma_{22} \end{pmatrix}.$$

Hence, we have

$$\begin{pmatrix} I & -S \\ 0 & I \end{pmatrix} \begin{pmatrix} \Sigma_{11} - \Sigma_{12}S^T & \Sigma_{12} \\ 0 & \Sigma_{22} \end{pmatrix} = \begin{pmatrix} \Sigma_{11} - S\Sigma_{21} & 0 \\ 0 & \Sigma_{22} \end{pmatrix}.$$

(b) From the results in (a) we find

$$\Sigma = \begin{pmatrix} I & S \\ 0 & I \end{pmatrix} \begin{pmatrix} \Sigma_{11} - S\Sigma_{21} & 0 \\ 0 & \Sigma_{22} \end{pmatrix} \begin{pmatrix} I & 0 \\ S^T & I \end{pmatrix}.$$

Therefore, using the matrix identity $(ABC)^{-1} = C^{-1}B^{-1}A^{-1}$, we have

$$\Sigma^{-1} = \begin{pmatrix} I & 0 \\ -S^T & I \end{pmatrix} \begin{pmatrix} \widetilde{\Sigma}^{-1} & 0 \\ 0 & \Sigma_{22}^{-1} \end{pmatrix} \begin{pmatrix} I & -S \\ 0 & I \end{pmatrix}.$$

Since

$$\begin{pmatrix} I & -S \\ 0 & I \end{pmatrix} \begin{pmatrix} u \\ v \end{pmatrix} = \begin{pmatrix} u - Sv \\ v \end{pmatrix},$$

we have

$$(u^T, v^T)\Sigma^{-1} \begin{pmatrix} u \\ v \end{pmatrix} = (u^T - v^TS^T, v^T) \begin{pmatrix} \widetilde{\Sigma}^{-1} & 0 \\ 0 & \Sigma_{22}^{-1} \end{pmatrix} \begin{pmatrix} u - Sv \\ v \end{pmatrix}$$

$$= (u^T - v^TS^T)\widetilde{\Sigma}^{-1}(u - Sv) + v^T\Sigma_{22}^{-1}v.$$

(c) From part (b) we know that

$$(x^T - \mu_1^T, y^T - \mu_2^T)\Sigma^{-1} \begin{pmatrix} x - \mu_1 \\ y - \mu_2 \end{pmatrix} = (x - \tilde{\mu})^T\widetilde{\Sigma}^{-1}(x - \tilde{\mu}) + (y - \mu_2)^T\Sigma_{22}^{-1}(y - \mu_2).$$

Hence,

$$f(x\,|\,y) = \frac{f(x, y)}{f(y)} \propto f(x, y)$$

$$= c_1 \exp\left[-\frac{1}{2}(x - \tilde{\mu})^T\widetilde{\Sigma}^{-1}(x - \tilde{\mu}) - \frac{1}{2}(y - \mu_2)^T\Sigma_{22}^{-1}(y - \mu_2) \right]$$

$$\propto \exp\left[-\frac{1}{2}(x - \tilde{\mu})^T\widetilde{\Sigma}^{-1}(x - \tilde{\mu}) \right],$$

from which the result follows.